Palle Thoft-Christensen
Michael J. Baker

Structural Reliability Theory and Its Applications

With 107 Figures

Springer-Verlag Berlin Heidelberg New York 1982

PALLE THOFT-CHRISTENSEN, Professor, Ph. D.
Institute of Building Technology
and Structural Engineering
Aalborg University Centre
Aalborg, Denmark

MICHAEL J. BAKER, B.Sc. (Eng)
Department of Civil Engineering
Imperial College of Science and Technology
London, England

Library of Congress Cataloging in Publication Data

Thoft-Christensen, Palle.
Structural reliability theory and its applications.

Includes bibliographies and index.
1. Structural stability. 2. Reliability (Engineering) I. Baker, Michael J. (Michael John), 1940- . II. Title.
TA656.5.T47 1982 624.1'71 82-10277
ISBN-13:978-3-642-68699-3(U.S.)

ISBN-13:978-3-642-68699-3 e-ISBN-13:978-3-642-68697-9
DOI: 10.1007/978-3-642-68697-9

2061/3020 - 543210

PREFACE

Structural reliability theory is concerned with the rational treatment of uncertainties in structural engineering and with the methods for assessing the safety and serviceability of civil engineering and other structures. It is a subject which has grown rapidly during the last decade and has evolved from being a topic for academic research to a set of well-developed or developing methodologies with a wide range of practical applications.

Uncertainties exist in most areas of civil and structural engineering and rational design decisions cannot be made without modelling them and taking them into account. Many structural engineers are shielded from having to think about such problems, at least when designing simple structures, because of the prescriptive and essentially deterministic nature of most codes of practice. This is an undesirable situation. Most loads and other structural design parameters are rarely known with certainty and should be regarded as random variables or stochastic processes, even if in design calculations they are eventually treated as deterministic. Some problems such as the analysis of load combinations cannot even be formulated without recourse to probabilistic reasoning.

There is a need for all structural engineers to develop an understanding of structural reliability theory and for this to be applied in design and construction, either indirectly through codes or by direct application in the case of special structures having large failure consequences, the aim in both cases being to achieve economy together with an appropriate degree of safety. The subject is now sufficiently well developed for it to be included as a formal part of the training of all civil and structural engineers, both at undergraduate and post-graduate levels. Courses on structural safety have been given at some universities for a number of years.

In writing this book we have tried to bring together under one cover the major components of structural reliability theory with the aim of making it possible for a newcomer to see and study the subject as a whole. The book should be of value to those with no prior knowledge of reliability theory, but it should also be of interest to those engineers involved in the development of structural and loading codes and to those concerned with the safety assessment of complex structures. The book does not try to cover all aspects of structural safety and no attempt is made, for example, to discuss structural failures except in general statistical terms.

It was the intention to make this book moderately self-contained and for this reason chapter 2 is devoted to the essential fundamentals of probability theory. However, readers who have had no training in this branch of mathematics would be well advised to study a more general text in addition. Topics such as the statistical theory of extremes, methods of parameter estimation and stochastic process theory are introduced in later chapters as and when they are required. The main core of the book is devoted to the so-called level 2 methods of analysis which have provided the key to fast computational procedures for structural reliability calculations. Other chapters cover the reliability of structural systems, load combinations, gross errors and some major areas of application.

The work is set out in the form of a textbook with a number of examples and simple exercises. The purpose of these is to illustrate the important principles and methods and to extend the scope of the main text with economy of space. The reader is warned against a too literal interpretation of some of the simple examples as these were *not* included to provide insight into particular practical problems. In some examples, the parameters of the probability distributions used in the calculations have been chosen quite arbitrarily or in such a way as to demonstrate the calculation procedure with maximum effect. This does not mean that the practical aspects of structural reliability theory have been overlooked - indeed, the theory would be of little value if it could not be applied. Chapter 11, on the application of reliability theory to the development of level 1 codes, attempts to address many of the practical problems faced by code writers in the selection of partial coefficients (partial factors); and in chapter 3, the modelling of load and resistance variables has been approached with applications strongly in mind. However, a complete book would be required to cover this subject in depth. Chapter 12 on offshore structures should be of interest to those working in this field.

In compiling the bibliography our approach has been to list only a selection of the more important works in each subject area, along with other works to which specific reference is made. Whilst many important contributions to the literature are thus omitted, it is considered that this selective approach will be of more help to the new reader.

We should like to acknowledge the major contributions to the field of structural reliability theory that have been made by a relatively small number of people, mainly during the last 10 to 15 years, and without which this book would not have been possible. The subject has benefitted from a large degree of international co-operation which has been stimulated by various bodies - in particular, the Joint Committee on Structural Safety under the chairmanship of J. Ferry Borges. The responsibility for this book must, however, rest with the authors and we should be pleased to receive notification of corrections or omissions of any nature.

Thanks are due to our respective colleagues in Aalborg and London for their helpful comments and contributions and in particular to Mrs. Kirsten Aakjær and Mrs. Norma Hornung who have undertaken the type-setting and drawing of figures, respectively, with such skill and efficiency.

We conclude with some words of caution. Structural reliability theory should not be thought of as the solution to all safety problems or as a procedure which can be applied in a mechanical fashion. In the right hands it is a powerful tool to aid decision making in matters of structural safety, but like other tools it can be misused. It should not be thought of as an alternative to more traditional methods of safety analysis, because *all* the information that is currently used in other approaches can and should be incorporated within a reliability analysis. On occasions the theory may give results which seem to contradict »experience». In this case, either »experience» will be found to have been incorrectly interpreted or some part of the reliability analysis will be at fault, generally the modelling. The resolution of these real or apparent contradictions will often provide considerable insight into the nature of the problem being examined, which can only be of benefit.

March, 1982

Palle Thoft-Christensen
Institute of Building Technology
and Structural Engineering
Aalborg University Centre
Aalborg, Denmark

Michael J. Baker
Department of Civil Engineering
Imperial College of Science and Technology
London, England

CONTENTS

Chapter 1

THE TREATMENT OF UNCERTAINTIES IN STRUCTURAL ENGINEERING

1.1 INTRODUCTION

Until fairly recently there has been a tendency for structural engineering to be dominated by deterministic thinking, characterised in design calculations by the use of specified minimum material properties, specified load intensities and by prescribed procedures for computing stresses and deflections. This deterministic approach has almost certainly been reinforced by the very large extent to which structural engineering design is codified and the lack of feedback about the actual performance of structures. For example, actual stresses are rarely known, deflections are rarely observed or monitored, and since most structures do not collapse the real reserves of strengths are generally not known. In contrast, in the field of hydraulic systems, much more is known about the actual performance of, say, pipe networks, weirs, spillways etc., as their performance in service can be relatively easily observed or determined.

The lack of information about the actual behaviour of structures combined with the use of codes embodying relatively high safety factors can lead to the view, still held by some engineers as well as by some members of the general public, that absolute safety can be achieved. Absolute safety is of course unobtainable; and such a goal is also undesirable, since absolute safety could be achieved only by deploying infinite resources.

It is now widely recognised, however, that some risk of unacceptable structural performance must be tolerated. The main object of structural design is therefore to ensure, at an acceptable level of probability, that each structure will not become unfit for its intended purpose at any time during its specified design life. Most structures, however, have multiple performance requirements, commonly expressed in terms of a set of serviceability and ultimate limit states, most of which are not independent; and thus the problem is much more complex than the specification of just a single probability.

1.1.1 Current Risk Levels

As a means of assessing the relative importance of structural failures as a cause of death, some comparative statistics for the U.K. are given in table 1.1 for a number of causes. These figures show that, at least for a typical Western European country, the risk to life from structural failures is negligible. For the 3 year period reported, the average number of deaths per annum directly attributable to structural failure was 14, divided almost equally between failures occurring during construction and the failures of completed structures. Other structural failures occur in which there are no deaths or personal injuries; but data on such failures are more difficult to assemble because in many countries they do not have to be reported.

In comparing the relative risks given in table 1.1, account should be taken of differences in exposure time typical for the various activities. For example, although air travel is associated with a high risk per hour, a typical passenger may be exposed for between only, say, 10 - 100 hours per year, leading to a risk of death of between 10^{-5} and 10^{-4} per year, (i.e. between 1 in 10^5 and 1 in 10^4). In contrast, most people spend at least 70% of their life indoors and are therefore exposed to the possible effects of structural failure; but this leads to an average annual risk per person of only 10^{-7}. Nevertheless, the only fair basis for comparing this risk is comparison with the inescapable minimum risk that has to be accepted by an individual member of society as beyond his control and for which no blame can be attributed to other people - for example, the risk of death by disease. Many people accept voluntary risks many orders of magnitude higher, but these should not be taken into account when considering structural safety.

Activity/Cause	Number of deaths per hour per 10^8 persons
Mountaineering (International)	2700
Air travel (International)	120
Deep water trawling	59
Car travel	56
Coal mining	21
Construction sites	7.7
Manufacturing	2.0
Accidents at home (all)	2.1
Accidents at home (able-bodied persons)	0.7
Fire at home	0.1
STRUCTURAL FAILURES	0.002
Natural causes (average, all ages)	129
Males aged 30 (all causes)	15
Females aged 30 (all causes)	13
Males aged 50 (all causes)	84
Females aged 50 (all causes)	51

Table 1.1 Comparative death risk. [Average 1970-1973 in U.K. based on Central Statistical Office, Abstract 1974].

In assessing the importance of structural failures, account should also be taken of the economic consequences of collapse and unserviceability. In fact the economic aspects may be regarded as dominant since the marginal returns in terms of lives saved for each additional £ 1 million invested in improving the safety of structures may be small in comparison with the benefits of investing the same sum in, say, road safety or health care. However, structures should, where possible, be designed in such a way that there is ample warning of impending failure and with brittle failure modes having larger safety margins than ductile modes (i.e. higher reliabilities).

1.1.2 Structural Codes

Most structural design is undertaken in accordance with codes of practice, which in many countries have legal status; although in the U.K., structural codes for buildings are simply »deemed to satisfy» the building regulations, meaning that compliance with the code automatically ensures compliance with the relevant clauses of the building laws. Structural codes typically and properly have a deterministic format and describe what are considered to be the minimum standards for design, construction and workmanship for each type of structure. Most codes can be seen to be evolutionary in nature, with changes being introduced or major revisions made at intervals of 3 - 10 years to allow for: new types of structural form (e.g. reinforced masonry), the effects of improved understanding of structural behaviour (e.g. of stiffened plated structures), the effects of changes in manufacturing tolerances or quality control procedures, a better knowledge of loads, etc.

Until recently, structural codes could be considered to be documents in which current good practice was codified; and these documents could be relied upon to produce safe, if not economic, structures. These high standards of safety were achieved for the majority of structures, not from an understanding of all the uncertainties that affect the loading and response, but by codifying practice that was known by experience to be satisfactory. The recent generation of structural codes, including the Euro-codes and the associated model codes for steel and concrete, are however more scientific in nature. They typically cover a wider range of structural elements and incorporate the results of much experimental and theoretical research. They are also more complex documents to assimilate and to use and the associated design costs are consequently higher, as are the risks of errors in interpretation.

The benefits of these new codes must therefore lie in the possibility of:

- increased overall safety for the same construction costs;
- the same or more consistent levels of safety with reduced construction costs;
- or, a combination of these two.

A further aim should be the trend, where appropriate, towards design procedures which can be applied with confidence to completely new forms of construction without the prior need for prototype testing.

The aims and benefits described above can only be achieved by a rational assessment of the various uncertainties associated with each type of structure and a study of their interactions. This is the essence of structural reliability analysis - the fundamentals of which will be described in the following chapters, along with some recent applications.

1.2 UNCERTAINTY

1.2.1 General

Structural reliability analysis is concerned with the rational treatment of uncertainties in structural engineering design and the associated problems of rational decision making. Consider the following statement:

- All quantities (except physical and mathematical constants) that currently enter into engineering calculations are in reality associated with some uncertainty. This fact has been implicitly recognised in current and previous codes. If this were not the case, a »safety factor» only slightly in excess of unity would suffice in all circumstances. The determination of appropriate standards of safety requires the quantification of these uncertainties by some appropriate means and a study of their interaction for the structure under consideration.

Before continuing, it is worth noting that the argument is sometimes advanced that the magnitudes of all variables are either bounded or can be restricted within specified limits by exercising appropriate standards of control, and that these bounding values should be used as the basis for design. In structural engineering, however, such arguments are inappropriate for a number of reasons:

- upper limits to individual loads and lower limits to material strength are not easily identified in practice (e.g. building occupancy loads, wind loads, the yield stress of steel, the cube or cylinder strength of concrete);
- even if such natural limits exist, their direct use in design is likely to be extremely uneconomic;
- limits imposed by quality control and testing can never be completely effective, particularly in the case of properties which can be measured only by destructive tests or in circumstances in which changes in the potential properties take place between sampling and use of the material (e.g. concrete);
- even if recognisable limits do exist, their use may not always be rational.

: **Example 1.1.** Consider a column supporting n floors of a building on which the loads are
: known to vary independently with time. Assuming that the load on each floor is physi-
: cally restricted by some hypothetical fail-safe device so that under no circumstances can
: it exceed some specified maximum value, and given that each load stays at this maximum
: value for, say, 1% of the time, the rational design load for the column can generally be
: shown to be less than the sum of the maximum loads. This design load will, of course, de-
: pend on the number of storeys supported and the design life of the structure.

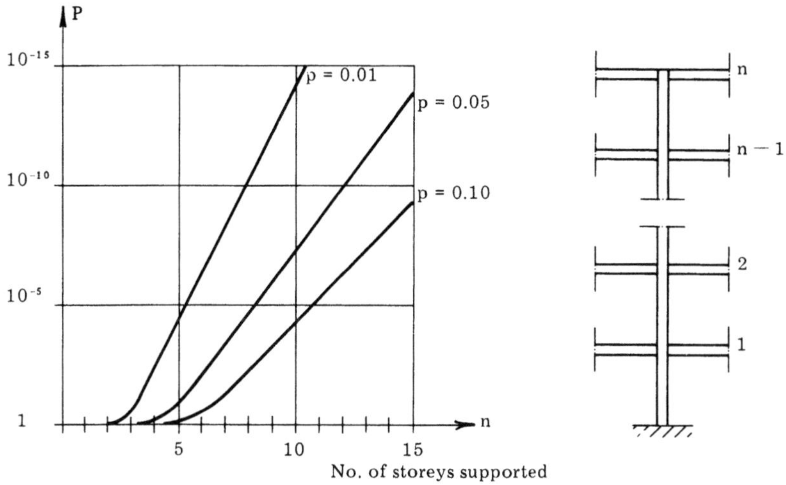

Figure 1.1

Figure 1.1 shows the probabilities P that the maximum column load at ground level will reach the sum of the maxima of the individual floor loads (i.e. the maximum possible column load) at some time during a 50 year period, on the assumption that the floor loads are mutually independent, that they remain constant for an hour and then change to some new random value, and that each has a 1% chance (p = 0.01) of being at its maximum value after each renewal (i.e. each floor is loaded to its maximum value for approximately 1% of the time). The figure shows that even with only six floors, the probability that the maximum possible column load occurs is as small as 10^{-6} in 50 years. Even if each floor is loaded to its maximum value for about 10% of the time (p = 0.1), the probability of the maximum possible column load occurring is still very small if the number of floors supported is 10 or more.

In such cases, it would be irrational and uneconomic to design for this worst possible condition. However, it should be noted that in practice the degree of conservatism depends on whether the individual floor loads are in fact independent; and the acceptable risk level is governed by the consequences of failure and the rapidity with which failure occurs. Some knowledge of the probabilities of occurrence of loads less than the maximum would also be required for each floor for a rational design.

The preceding example demonstrates that although there is no (strictly, negligible) uncertainty in the magnitude of the maximum load intensity on each floor, there may be appreciable uncertainty in the magnitude of the maximum combined load carried by the column.

We return now to the question of classifying the various types of uncertainty that can arise in structural reliability analysis. However, before these are discussed it is helpful to introduce the concept of basic variables. This concept is treated in more detail in chapter 5.

1.2.2 Basic Variables

For the purposes of quantifying uncertainties in the field of structural engineering and for subsequent reliability analysis it is necessary to define a set of *basic variables*. These are defined as the set of basic quantities governing the static or dynamic response of the structure. Basic variables are quantities such as mechanical properties of materials, dimensions, unit weights, environmental loads, etc. They are basic in the sense that they are the most fundamental quantities normally recognised and used by designers and analysts in structural calculations.

Thus, the yield stress of steel can be considered as a basic variable, although this property is itself dependent on chemical composition and various micro-structural parameters. Mathematical models involving these latter parameters are often used by steel producers for predicting the mechanical properties of structural steels and for the purposes of quality control. However, it is generally sensible to treat the mechanical properties as basic variables for the purposes of structural reliability analysis. One justification is that more statistical data are available for the mechanical properties of, say, steels than for the more basic metallurgical properties.

It should also be mentioned that it is generally impracticable to try to obtain sufficient statistical data to model the variations in the strength of complete structural components directly. Reliance must be placed on the ability of the analyst to synthesise this higher level information when required.

Ideally, basic variables should be chosen so that they are statistically independent quantities. However, this may not always be possible if the strength of a structure is known to be dependent on, for example, any two mechanical properties that are known to be correlated, e.g. the tensile strength and the compressive strength of a batch of concrete.

1.2.3 Types of Uncertainty

For the purposes of structural reliability analysis it is necessary to distinguish between at least three types of uncertainty - physical uncertainty, statistical uncertainty and model uncertainty. These are now described.

It should be noted that in the following, random variables will be denoted by upper case letters, e.g. X, and their outcomes by lower case, e.g. x.

Physical uncertainty: Whether or not a structure or structural element fails when loaded depends in part on the actual values of the relevant material properties that govern its strength. The reliability analyst must therefore be concerned with the nature of the actual variability of physical quantities, such as loads, material properties and dimensions. This variability can be described in terms of probability distributions or stochastic processes and some typical examples are discussed in detail in chapter 3. However, physical variability can be quantified only by examining sample data; but, since sample sizes are limited by practical and economic considerations, some uncertainty must remain. This practical limit gives rise to so-called statistical uncertainty.

Statistical uncertainty: As will be discussed in later chapters, statistics, as opposed to probability, is concerned with inference, and in particular with the inferences that can be drawn from sample observations. Data may be collected for the purposes of building a probabilistic model of the physical variability of some quantity which will involve, firstly the selection of an appropriate probability distribution type, and then determination of numerical values for its parameters. Common probability distributions have between one and four parameters which immediately places a lower bound on the sample size required, but in practice very large samples are required to establish reliable estimates of the numerical values of parameters. For a given set of data, therefore, the distribution parameters may themselves be considered to be random variables, the uncertainty in which is dependent on the amount of sample data - or, in general, on the amount of data and any prior knowledge. This uncertainty is termed *statistical uncertainty* and, unlike physical variability, arises solely as a result of lack of information.

Model uncertainty: Structural design and analysis make use of mathematical models relating desired output quantities (e.g. the deflection at the centre of a reinforced concrete beam) to the values of a set of input quantities or basic variables (e.g. load intensities, modulus of elasticity, duration of loading, etc.). These models are generally deterministic in form (e.g. linear-elastic structural analysis) although they may be probabilistic (e.g. calculation of the peak response of an offshore structure to stochastic wave loading). Furthermore, they may be based on an intimate understanding of the mechanics of the problem (e.g. plastic collapse analysis of a steel portal frame) or they may be highly empirical (e.g. punching shear at tubular joint connections in offshore jacket structures). However, with very few exceptions, it is rarely possible to make highly accurate predictions about the magnitude of the response of typical civil engineering structures to loading even when the governing input quantities are known exactly. In other words, the response of typical structures and structural elements contains a component of uncertainty in addition to those components arising from uncertainties in the values of the basic loading and strength variables. This additional source of uncertainty is termed *model uncertainty* and occurs as a result of simplifying assumptions, unknown boundary conditions and as a result of the unknown effects of other variables and their interactions which are not included in the model. For example, the shear strength of nominally similar reinforced concrete beams exhibits considerable scatter even when due allowance has been made for the various known differences between test specimens.

The model uncertainty associated with a particular mathematical model may be expressed in terms of the probability distribution of a variable X_m defined as

$$X_m = \frac{\text{actual strength (response)}}{\text{predicted strength (response) using model}} \tag{1.1}$$

In many components and structures, model uncertainties have a large effect on structural reliability and should not be neglected.

1.3 STRUCTURAL RELIABILITY ANALYSIS AND SAFETY CHECKING

The preceding remarks were concerned with the various types and sources of uncertainty that need to be taken into account in predicting the ability of a structure to withstand the actual but unknown loads that will be applied to it. We shall now consider the various ways in which these predictions can be made and the use that can be made of them. But first some definitions and preliminaries.

1.3.1 Structural Reliability

The term *structural reliability* should be considered as having two meanings - a general one and a mathematical one.

- In the most general sense, the *reliability* of a structure is its ability to fulfil its design purpose for some specified time.
- In a narrow sense it is the *probability* that a structure will not attain each specified limit state (ultimate or serviceability) during a specified *reference period.*

In this book we shall be concerned with structural reliability in the narrow sense and shall generally be treating each limit state or failure mode separately and explicitly. However, most structures and structural elements have a number of possible failure modes, and in determining the overall reliability of a structural system this must be taken into account, making due allowance for the correlations arising from common sources of loading and common material properties. These aspects of the problem are covered in chapters 7 and 8.

However, although the definition above may seem clear, it is necessary to examine with care exactly what is meant by »the probability that a structure will not attain each specified limit state during a specified reference period».

Consider first the need for defining a *reference period.* Because the majority of structural loads vary with time in an uncertain manner, the probability that any selected load intensity will be exceeded in a fixed interval of time is a function of the length of that interval (and possibly the time at which it begins). Hence, in general, structural reliability is dependent on time of exposure to the loading environment. It is also affected if material properties change with time. Only for the rare cases, when loads and strength are constant, can the reference period be ignored. In such cases the loads are applied once and the structure either does or does not fail (e.g. when the structure or component is loaded entirely by its own self weight).

The second question is more demanding. What is meant by »the probability that . . .»? This is best explored by a simple example.

: **Example 1.2.** Assume that an offshore structure is idealised as a uniform vertical cantilever rigidly connected to the sea bed. The structure will fail when the moment S induced at the root of the cantilever exceeds the flexural strength R. Assume further that R and S are random variables whose statistical distributions are known very precisely as a result of a very long series of measurements. R is a variable representing the variations in strength between nominally identical structures, whereas S represents the maximum load effects in successive T year periods. The distributions of R and S are both assumed to be stationary with time. Under these assumptions, the probability that the structure will collapse during any reference period of duration T years will be shown in chapter 4 to be given by

$$P_f = P(M \leqslant 0) = \int_{-\infty}^{+\infty} F_R(x)\, f_S(x) dx \tag{1.2}$$

: where

: $$M = R - S \qquad (1.3)$$
:
: and where F_R is the probability distribution function of R and f_S the probability density
: function of S. These terms are defined more fully in chapter 2.

Because of the definition of R and S in terms of frequentist probabilities, the probability determined from equation (1.2) may be interpreted as a long-run failure frequency. Similarly the reliability $\mathcal{R}$, defined as

$$\mathcal{R} = 1 - P_f \qquad (1.4)$$

may be interpreted as a long-run survival frequency or long-run reliability and is the percentage of a notionally infinite set of nominally identical structures which survive for the duration of the reference period T. $\mathcal{R}$ may therefore be called a *frequentist reliability*. If, however, we are forced to focus our attention on one particular structure (and this is generally the case for »one-off» civil engineering structures), $\mathcal{R}$ may also be interpreted as a measure of the reliability of that particular structure.

This interpretation of reliability is fundamentally different from that given above, because, although the structure may be sampled at random from the theoretically infinite population described by the random variable R, once the particular structure has been selected (and, in practice, constructed) the reliability becomes the probability that the fixed, but unknown, resistance r will be exceeded by the as yet »un-sampled» reference period extreme load effect S [note that lower case r is used here to denote the outcome of R]. The numerical value of the failure probability remains the same but is now dependent upon two radically different types of uncertainty - firstly, the physical variability of the extreme load effect, and, secondly, lack of knowledge about the true value of the fixed but unknown resistance. This type of probability does not have a relative frequency interpretation and is commonly called a subjective probability. The associated reliability can be called a subjective or *Bayesian reliability*. For a particular structure, the numerical value of this reliability changes as the state of knowledge about the structure changes - for example, if non-destructive tests were to be carried out on the structure to estimate the magnitude of r. In the limit when r becomes known exactly, the probability of failure given by equation (1.2) changes to

$$P_f = P(r - S \leqslant 0) = 1 - F_S(r) \qquad (1.5)$$

This special case may also be interpreted as a *conditional failure probability* with a relative frequency interpretation, i.e.

$$P_f = P(R - S \leqslant 0 \mid R = r) \qquad (1.6)$$

The symbol $\mid$ may be read as »given that».

1.3.2 Methods of Safety Checking

Many developments have taken place in the field of structural reliability analysis during the last 10 years and to the newcomer the literature may seem confusing. To help clarifying the situation, the Joint Committee on Structural Safety (an international body sponsored by such international organisations as CEB, CIB, CECM, IABSE, IASS, FIP and RILEM) set up a subcommittee in 1975 to provide a broad classification system for the different methods then being proposed for checking the safety of structures and to establish the main differences between them. This classification is still useful.

Methods of structural reliability analysis can be divided into two broad classes. These are:

Level 3: Methods in which calculations are made to determine the »exact» probability of failure for a structure or structural component, making use of a full probabilistic description of the joint occurrence of the various quantities which affect the response of the structure and taking into account the true nature of the failure domain.

Level 2: Methods involving certain approximate iterative calculation procedures to obtain an approximation to the failure probability of a structure or structural system, generally requiring an idealisation of failure domain and often associated with a simplified representation of the joint probability distribution of the variables.

In theory, both level 3 and level 2 methods can be used for checking the safety of a design or directly in the design process, provided a target reliability or reliability index has been specified.

For the sake of completeness, some mention should also be made of level 1 methods at this stage. These are not methods of reliability analysis, but are methods of design or safety checking.

Level 1: Design methods in which appropriate degrees of structural reliability are provided on a structural element basis (occasionally on a structural basis) by the use of a number of partial safety factors, or partial coefficients, related to pre-defined characteristic or nominal values of the major structural and loading variables.

A *level 1 structural design*, with the explicit consideration of a number of separate limit states, is what is now commonly called *limit-state design*. It is preferable, however, that this should be called *level 1 design*, and that the term limit-state should be used solely to describe the separate limiting performance requirements.

The terms, level 1, 2 and 3 will be discussed in detail in chapters 5, 6, 7, 8 and 11. The fundamental distinctions between levels 2 and 3 cannot easily be understood at this stage until the necessary background in probability theory has been covered (chapter 2). However, the three levels of safety checking should be seen as a hierachy of methods in which level 2 methods are an approximation to level 3 methods and in which level 1 methods are a discretisation of level 2 methods (i.e. giving identical designs to level 2 methods for only a few discrete sets of values of the structural design parameters).

For practical purposes - for example, for direct use in design or for evaluating level 1 partial factors - it is necessary to have a method of reliability analysis which is computationally fast and efficient and which produces results with the desired degree of accuracy. The only methods which currently satisfy these requirements are the level 2 methods, although analysis by Monte-Carlo simulation is sometimes feasible. In this book emphasis is placed on the theory and application of level 2 methods and their use in the design of level 1 structural codes.

BIBLIOGRAPHY

[1.1] CIRIA: *Rationalisation of Safety and Serviceability Factors in Structural Codes.* Construction Industry Research and Information Association, Report No. 63, 1977.

[1.2] Cornell, C. A.: *Bayesian Statistical Decision Theory and Reliability-Based Design.* International Conference on Structural Safety and Reliability, Washington, 1969. (pub. Pergamon 1972).

[1.3] Ditlevsen, O.: *Uncertainty Modeling.* McGraw Hill, 1981.

[1.4] Freudenthal, A. M., Garretts, J. M. and Shinozuka, M.: *The Analysis of Structural Safety.* Journal of the Structural Division, ASCE, Vol. 92, No. ST1, Feb. 1966.

[1.5] Joint Committee on Structural Safety, CEB - CECM - CIB - FIP - IABSE - IASS - RILEM: *First Order Reliability Concepts for Design Codes.* CEB Bulletin No. 112, July 1976.

[1.6] Joint Committee on Structural Safety, CEB - CECM - CIB - FIP - IABSE - IASS - RILEM: *International System of Unified Standard Codes for Structures.* Volume I: Common Unified Rules for Different Types of Construction and Material. CEB/FIP 1978.

[1.7] Joint Committee on Structural Safety, CEB - CECM - CIB - FIP - IABSE - IASS - RILEM: *General Principles on Reliability for Structural Design.* International Association for Bridge and Structural Engineering, 1981.

[1.8] Leporati, E.: *The Assessment of Structural Safety.* Research Studies Press, 1979.

[1.9] Nordic Committee on Building Regulations: *Recommendation for Loading and Safety Regulations for Structural Design.* NKB-Report No. 36, Nov. 1978.

Chapter 2

FUNDAMENTALS OF PROBABILITY THEORY

2.1 INTRODUCTION

As emphasized in chapter 1, modern structural reliability analysis is based on a probabilistic point of view. It is therefore important to get a profound knowledge of at least some part of probability theory. It is beyond the scope of this book to give a thorough presentation of probability theory on a rigorous axiomatic basis. What is needed to understand the following chapters is some knowledge of the fundamental assumptions of modern probability theory, combined with a number of definitions and theorems. It will be assumed that the reader is familiar with the terminology and the algebra of simple set theory.

The purpose of this chapter is therefore to give a self-contained presentation of probability theory with emphasis on concepts of importance for structural reliability analysis.

2.2 SAMPLE SPACE

A standard way of determining the yield stress of a material such as steel is to perform a number of simple tensile tests with specimens made from the material in question. By each test a value for the yield stress is determined but this value will probably be different from test to test. Therefore, in this connection, the yield stress must be taken as an uncertain quantity and it is in accordance with this point of view said to be a *random* quantity. The set of all possible outcomes of such tests is called the *sample space* and each individual outcome is a *sample point*. The sample space for the yield stress is the open interval $]0 ; \infty[$ - that is the set of all positive real numbers. Note that this sample space has an infinite number of sample points. It is an example of a *continuous* sample space. A sample space can also be *discrete*, namely when the sample points are discrete and countable entities.

Example 2.1. Consider a simply supported beam AB with two concentrated forces P_1 and P_2 as shown on figure 2.1. Further, let the possible values of P_1 and P_2 be 4, 5, 6, and 3, 4, respectively. In this example all values are in kN. The sample space for the loading will then be the set

$$\Omega = \{(4, 3), (4, 4), (5, 3), (5, 4), (6, 3), (6, 4)\} \quad (2.1)$$

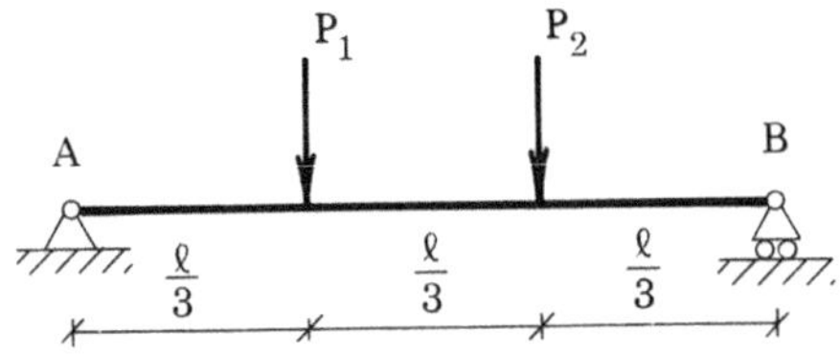

Figure 2.1

This sample space is discrete. Further, it has a finite number of sample points. Therefore, it is called a *finite* sample space. A sample space with the countable infinite number of sample points is called an *infinite* sample space.

Note that the sample spaces for the loads P_1 and P_2 are $\Omega_1 = \{4, 5, 6\}$ and $\Omega_2 = \{3, 4\}$, respectively. Also note that $\Omega = \Omega_1 \times \Omega_2$. Show as an exercise that the sample space for the reaction R_A in point A is $\Omega_A = \{11/3,\ 12/3,\ 13/3,\ 14/3,\ 15/3,\ 16/3\}$.

A subset of a sample space is called an *event*. An event is therefore a set of sample points. If it contains no sample points it is called an *impossible event*. A *certain event* contains all the sample points in the sample space - that is, a certain event is equal to the sample space itself.

Example 2.2. Consider again the beam in figure 2.1. The sample space for the reaction R_A is $\Omega_A = \{11/3,\ 12/3,\ 13/3,\ 14/3,\ 15/3,\ 16/3\}$. The subset $\{15/3,\ 16/3\}$ is the event that R_A is equal to 15/3 or 16/3.

Let E_1 and E_2 be two events. The *union* of E_1 and E_2 is an event denoted by $E_1 \cup E_2$ and it is the subset of sample points that belong to E_1 and/or E_2. The *intersection* of E_1 and E_2 is an event denoted by $E_1 \cap E_2$ and is the subset of sample points belonging to both E_1 and E_2. The two events E_1 and E_2 are said to be *mutually exclusive* if they are disjoint, that is if they have no sample points in common. In this case $E_1 \cap E_2 = \emptyset$, where $\emptyset$ is the impossible event (an empty set).

Let Ω be a sample space and E an event. The event containing all the sample points in Ω that are not in E is called the *complementary event* and is denoted by $\overline{E}$. Obviously, $E \cup \overline{E} = \Omega$ and $E \cap \overline{E} = \emptyset$.

It is easy to show that the intersection and union operations obey the following commutative, associative, and distributive laws

$$E_1 \cap E_2 = E_2 \cap E_1 \quad \text{and} \quad E_1 \cup E_2 = E_2 \cup E_1 \tag{2.2}$$

$$\left.\begin{aligned} E_1 \cap (E_2 \cap E_3) &= (E_1 \cap E_2) \cap E_3 \\ E_1 \cup (E_2 \cup E_3) &= (E_1 \cup E_2) \cup E_3 \end{aligned}\right\} \tag{2.3}$$

$$\left.\begin{aligned} E_1 \cap (E_2 \cup E_3) &= (E_1 \cap E_2) \cup (E_1 \cap E_3) \\ E_1 \cup (E_2 \cap E_3) &= (E_1 \cup E_2) \cap (E_1 \cup E_3) \end{aligned}\right\} \tag{2.4}$$

Due to these laws it makes sense to consider the intersection or the union of the events $E_1, E_2, \ldots, E_n$. These new events are denoted

$$\bigcap_{i=1}^{n} E_i = E_1 \cap E_2 \cap \ldots \cap E_n \tag{2.5}$$

and

$$\bigcup_{i=1}^{n} E_i = E_1 \cup E_2 \cup \ldots \cup E_n \tag{2.6}$$

: **Exercise 2.1.** Prove the so-called *De Morgan's laws*

$$E_1 \cap E_2 = \overline{\overline{E}_1 \cup \overline{E}_2} \tag{2.7}$$

$$E_1 \cup E_2 = \overline{\overline{E}_1 \cap \overline{E}_2} \tag{2.8}$$

2.3 AXIOMS AND THEOREMS OF PROBABILITY THEORY

In this section is shown how a probability measure can be assigned to any event. Such a probability measure is a set function because an event is a subset of the sample space. Further, the probability of the certain event (the sample space itself) is unity. Finally, it is reasonable to assume that the probability of the union of mutually exclusive events is equal to the sum of the probability of the individual events. These assumptions are given a mathematical, precise formulation by the following fundamental axioms of probability theory.

Axiom 1

For any event E

$$0 \leqslant P(E) \leqslant 1 \tag{2.9}$$

where the function P is the *probability measure.* P(E) is the probability of the event E.

Axiom 2

Let the sample space be Ω. Then

$$P(\Omega) = 1 \tag{2.10}$$

Axiom 3

If $E_1, E_2, \ldots, E_n$ are mutually exclusive events then

$$P\left(\bigcup_{i=1}^{n} E_i\right) = \sum_{i=1}^{n} P(E_i) \tag{2.11}$$

Exercise 2.2. Prove the following theorems

$$P(\overline{E}) = 1 - P(E) \tag{2.12}$$

$$P(\emptyset) = 0 \tag{2.13}$$

$$P(E_1 \cup E_2) = P(E_1) + P(E_2) - P(E_1 \cap E_2) \tag{2.14}$$

Example 2.3. Consider the statically determinate structural system with 7 elements shown in figure 2.2. Let the event that element »i» fails be denoted by F_i and let the probability of failure of element »i» be $P(F_i)$. Further assume that failures of the individual members are statistically independent, that is $P(F_i \cap F_j) = P(F_i) \cdot P(F_j)$ for any pair of (i, j). The failure of any member will result in system failure for this statically determinate structure. Thus,

$$P(\text{failure of structure}) = P(F_1 \cup \ldots \cup F_7) = P(\bigcup_{i=1}^{7} F_i)$$

$$= 1 - P(\overline{\bigcup_{i=1}^{7} F_i}) = 1 - P(\bigcap_{i=1}^{7} \overline{F}_i) \tag{2.15}$$

according to De Morgan's law (2.8). Because of statistical independence, (2.15) can be written

$$P(\text{failure of structure}) = 1 - P(\overline{F}_1) \cdot P(\overline{F}_2) \ldots P(\overline{F}_7)$$

$$= 1 - (1 - P(F_1))(1 - P(F_2)) \ldots (1 - P(F_7))$$

Let $P(F_1) = P(F_3) = P(F_5) = P(F_7) = 0.02$, $P(F_2) = P(F_6) = 0.01$, and $P(F_4) = 0.03$. Then

$$P(\text{failure of structure}) = 1 - 0.98^4 \cdot 0.99^2 \cdot 0.97 = 1 - 0.8769 = 0.1231$$

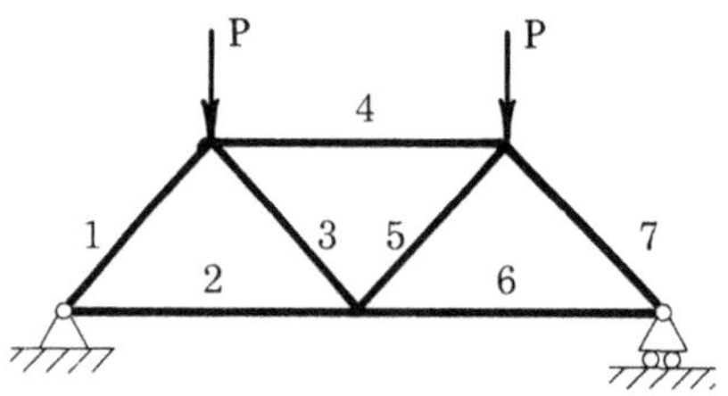

Figure 2.2

In many practical applications the probability of occurrence of event E_1 conditional upon the occurrence of event E_2 is of great interest. This probability called the *conditional probability* is denoted $P(E_1 \mid E_2)$ and is defined by

$$P(E_1 \mid E_2) = \frac{P(E_1 \cap E_2)}{P(E_2)} \tag{2.16}$$

if $P(E_2) > 0$. The conditional probability is not defined for $P(E_2) = 0$.

Event E_1 is said to be statistically *independent* of event E_2 if

$$P(E_1 | E_2) = P(E_1) \tag{2.17}$$

that is, if the occurrence of E_2 does not affect the probability of E_1.

From equation (2.16) the probability of the event $E_1 \cap E_2$ is given by

$$P(E_1 \cap E_2) = P(E_1 | E_2)P(E_2) = P(E_2 | E_1)P(E_1) \tag{2.18}$$

If E_1 and E_2 are statistically independent (2.18) becomes

$$P(E_1 \cap E_2) = P(E_1)P(E_2) \tag{2.19}$$

The rule (2.19) is called the *multiplication* rule and has already been used in example 2.3.

Exercise 2.3. Show that $\overline{E}_1$ and $\overline{E}_2$ are statistically independent, when E_1 and E_2 are statistically independent.

Exercise 2.4. Show that

$$P(E_1 \cap E_2 \cap E_3) = P(E_1 | E_2 \cap E_3)P(E_2 | E_3)P(E_3) \tag{2.20}$$

Example 2.4. Consider again the structure in figure 2.2. It is now assumed for the sake of simplicity, however, that only element 2 and 6 can fail. Therefore,

$$P(\text{failure of structure}) = P(F_2 \cup F_6) = P(F_2) + P(F_6) - P(F_2 \cap F_6)$$

$$= P(F_2) + P(F_6) - P(F_2 | F_6)P(F_6) \tag{2.21}$$

If F_2 and F_6 are statistically independent as in example 2.3 and if $P(F_2) = P(F_6) = 0.01$ then

$$P(\text{failure of structure}) = 0.01 + 0.01 - 0.01 \cdot 0.01 = 0.0199$$

But if F_2 and F_6 are not independent then knowledge of $P(F_2 | F_6)$ is required. If the two elements are fabricated from the same steel bar it is reasonable to expect them to have the same strength. Further, they have the same loading, and therefore in this special case, one can expect $P(F_2 | F_6)$ to be close to 1. With $P(F_2 | F_6) = 1$ one gets from (2.21)

$$P(\text{failure of structure}) = 0.01 + 0.01 - 1 \cdot 0.01 = 0.0100$$

Finally, the so-called *Bayes' theorem* will be derived. Let the sample space Ω be divided into n mutually exclusive events $E_1, E_2, \ldots, E_n$ (see figure 2.3, where n = 4). Let A be an event in the same sample space. Then

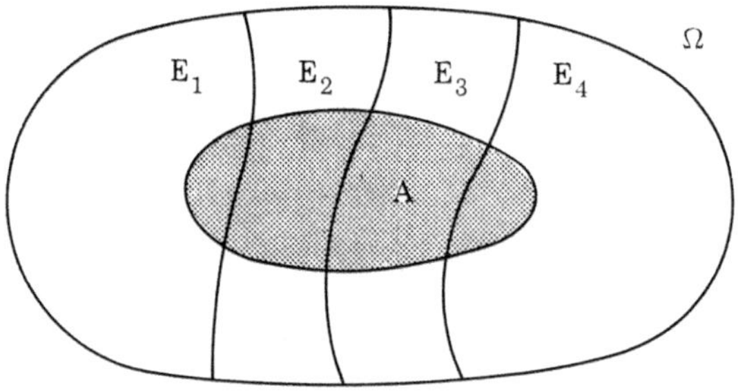

Figure 2.3

$$P(A) = P(A \cap E_1) + P(A \cap E_2) + \ldots + P(A \cap E_n)$$

$$= P(A \mid E_1)P(E_1) + P(A \mid E_2)P(E_2) + \ldots + P(A \mid E_n)P(E_n)$$

$$= \sum_{i=1}^{n} P(A \mid E_i)P(E_i) \tag{2.22}$$

From the definition (2.16) follows

$$P(A|E_i)\ P(E_i) = P(E_i|A)\ P(A)$$

so that

$$P(E_i \mid A) = \frac{P(A \mid E_i)P(E_i)}{P(A)} \tag{2.23}$$

or by using (2.22)

$$P(E_i \mid A) = \frac{P(A \mid E_i)P(E_i)}{\sum_{j=1}^{n} P(A \mid E_j)P(E_j)} \tag{2.24}$$

This is the important Bayes' theorem.

Example 2.5. Assume that a steel girder has to pass a given test before application. Further, assume from experience that 95% of all girders are found to pass the test, but the test is assumed only 90% reliable. Therefore, a conclusion based on such a test has a probability of 0.1 of being erroneous. The problem is now the following: What is the probability that a perfect girder will pass the test? Let E be the event that the girder is perfect and let A be the event that it passes the test.

$P(E \mid A) = 0.90$ and

$P(\overline{E} \mid \overline{A}) = 0.90$

so that

$$P(E|\bar{A}) = 1 - 0.90 = 0.10$$

From experience P(A) = 0.95. The problem is to find P(A|E). The events A and $\bar{A}$ are mutually exclusive, so that according to (2.22)

$$P(E) = P(E|A)P(A) + P(E|\bar{A})P(\bar{A}) = 0.90 \cdot 0.95 + 0.10 \cdot 0.05 = 0.860$$

Finally,

$$P(A|E) = \frac{P(E|A) \cdot P(A)}{P(E)} = \frac{0.90 \cdot 0.95}{0.86} = 0.994$$

Example 2.6. Consider a number of tensile specimens designed to support a load of 2 kN. The problem is now to estimate the probability that a specimen can support a load of 2.5 kN. Based on previous experiments it is estimated that there is a probability of 0.80 that a specimen can carry 2.5 kN. Further, it is known that 50% of those not able to support 2.5 kN fail at loads less than 2.3 kN.

The probability of 0.80 mentioned above can now be updated if the following test is successful. A single specimen is loaded to 2.3 kN.

Let E be the event that the specimen can support 2.5 kN and A the event that the test is successful (the specimens can support 2.3 kN). Then $P(\bar{A}|\bar{E}) = 0.5$, and P(E) = 0.80. Further P(A|E) = 1.0 so that Bayes' theorem gives

$$P(E|A) = \frac{P(A|E)P(E)}{P(A|E)P(E) + P(A|\bar{E})P(\bar{E})} = \frac{1.0 \cdot 0.80}{1.0 \cdot 0.80 + 0.5 \cdot 0.20} = 0.89$$

The previous value of 0.80 for the probability that a specimen can carry 2.5 kN is in this way updated to 0.89.

2.4 RANDOM VARIABLES

The outcome of experiments will in most cases be numerical values. But this will not always be true. If, for example, one wants to check whether a given structure can carry a given load the outcome may be yes or no. However, in such a case it is possible to assign a numerical value to the outcome, for example the number 1 to the event that the structure can carry the load, and the number 0 to the event that the structure cannot carry the load. Note that the numbers 0 and 1 are artificially assigned numerical values and therefore, other values could have been associated with the events in question. In this way it is possible to identify possible outcomes of a random phenomenon by numerical values. In most cases these values will simply be the outcomes of the phenomenon but as mentioned it may be necessary to assign the numerical values artificially.

In this way an outcome or event can be identified through the value of a function called a *random variable*. A random variable is a function which maps events in the sample space Ω into the real line R. Usually a random variable is denoted by a capital letter such as X. To emphasize the domain of X the random variable is often written $X : \Omega \curvearrowright R$. The concept of a continuous random variable is illustrated in figure 2.4. The event $E_1 \subset \Omega$, where Ω is a continuous sample

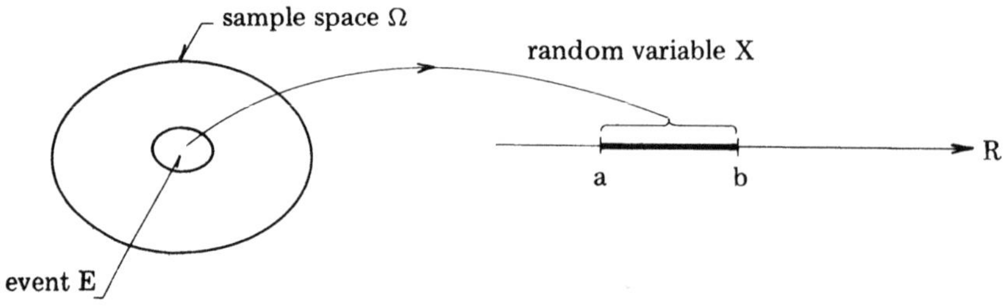

Figure 2.4

space, is mapped by the function X on to the interval $[a\,;\,b] \subset R$. If the sample space is discrete, the random variable is called a discrete random variable.

In section 2.3 the probability of an event E is introduced by the probability measure P. In this section, it is shown how a numerical value is associated with any event by the random variable. This permits a convenient analytical and graphical description of events and associated probabilities. Usually the argument ω in $X(\omega)$ is omitted. Similarly, the abbreviation $P(X \leqslant x)$ is used for $P(\{\omega : X(\omega) \leqslant x\})$.

First consider a *discrete random variable* X. This is a function that takes on only a finite or countably infinite number of discrete values. For such a random variable the *probability mass function* p_X is defined by

$$p_X(x) = P(X = x) \tag{2.25}$$

where X is the random variable, and $x = x_1, x_2, \ldots, x_n$, and where n can be finite or infinite. Note that different symbols are used for the random variable and its values, namely X and x, respectively. It is a direct consequence of the axioms (2.9) - (2.11) that

$$0 \leqslant p_X(x) \leqslant 1 \tag{2.26}$$

$$\sum_{i=1}^{n} p_X(x) = 1 \tag{2.27}$$

$$P(a \leqslant X \leqslant b) = \sum_{x_i \leqslant b} p_X(x_i) - \sum_{x_i \leqslant a} p_X(x_i) \tag{2.28}$$

The *probability distribution function* $P_X : R \to R$ is related to p_X by

$$P_X(x) = P(X \leqslant x) = \sum_{x_i \leqslant x} p_X(x_i) \tag{2.29}$$

By the definition (2.29) the value $P_X(x)$ is the probability of the event that the random variable X takes on values equal to or less than x.

Example 2.7. Consider again example 2.1 and let $P(P_1 = 4) = 0.3$, $P(P_1 = 5) = 0.5$ and $P(P_1 = 6) = 0.2$. The probability mass function p_{P_1} and the probability distribution function P_{P_1} for the random variable P_1 are shown on figure 2.5. Note that the circled points are not included in $P_{P_1}(x)$.

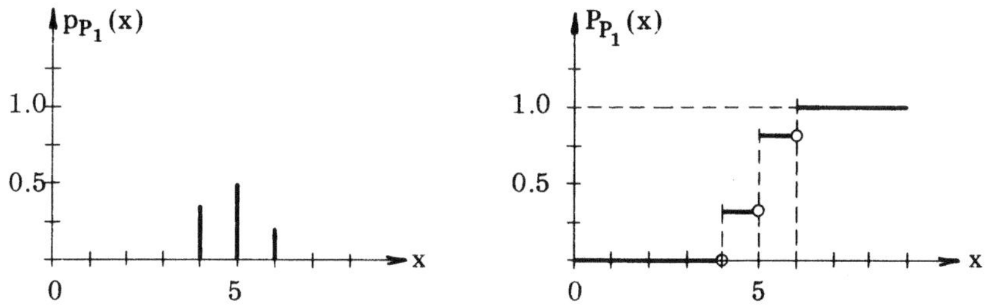

Figure 2.5

Next consider a *continuous random variable* X. This is a function which can take on any value within one or several intervals. For such a random variable the probability for it to assume a specific value is zero. Therefore, the probability mass function defined in (2.25) is of no interest. However, the *probability distribution function* $F_X : R \rightarrow R$ can still be defined by

$$F_X(x) = P(X \leqslant x) \quad , \quad x \in R \tag{2.30}$$

It is often useful to use the derivative probability function. This function is called the *probability density function* $f_X : R \rightarrow R$ and is defined by

$$f_X(x) = \frac{dF_X(x)}{dx} \tag{2.31}$$

assuming of course that the derivative exists. Note that the symbol $p_X(x)$ is used for the probability mass function and the symbol $f_X(x)$ for the probability density function.

Example 2.8. Figure 2.6 shows the probability density function f_X and the probability distribution function F_X for a continuous random variable X.

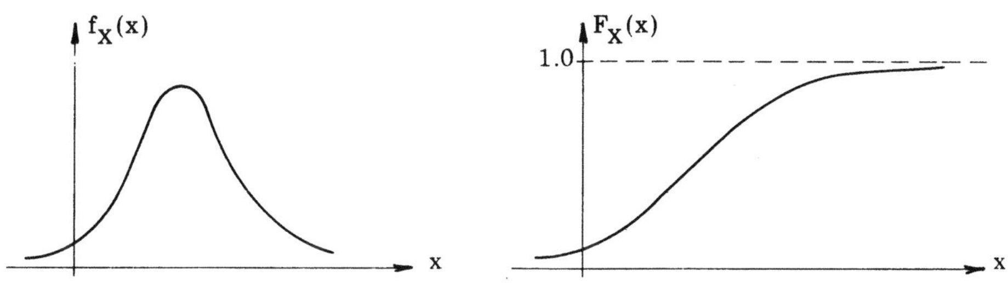

Figure 2.6

It follows directly from the axioms (2.9) - (2.11) that for any probability distribution function

(1) $F_X(-\infty) = 0$ and $F_X(\infty) = 1$

(2) F_X is non-decreasing

Inversion of the equation (2.31) gives

$$F_X(x) = \int_{-\infty}^{x} f_X(t)dt \tag{2.32}$$

for a continuous random variable. From (2.32), it follows that

$$\int_{-\infty}^{\infty} f_X(t)dt = F_X(\infty) = 1 \tag{2.33}$$

It is sometimes useful to use a *mixed continuous-discrete random variable*, i.e. a continuous random variable admitting a countable number of discrete values with a non-zero probability as shown in figure 2.7.

In this case the area under the curve in figure 2.7 is equal to $1 - 0.2 - 0.1 = 0.7$.

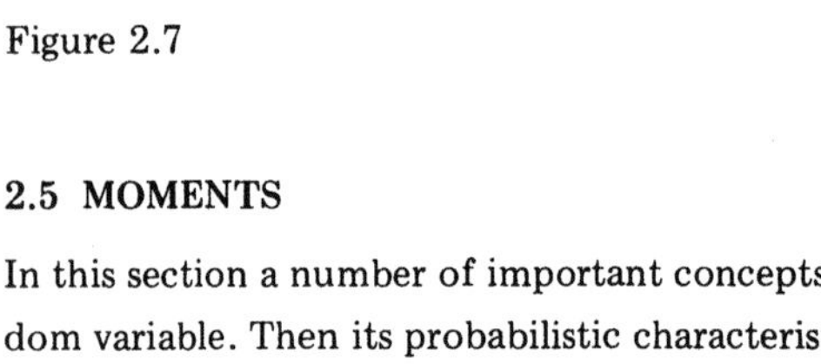

Figure 2.7

2.5 MOMENTS

In this section a number of important concepts will be introduced. Let X be a continuous random variable. Then its probabilistic characteristics are described by the distribution function F_X. However, in many applications the form of F_X is not known in all details. It is therefore useful to have an approximate description of a random variable stressing its most important features. When F_X (or f_X) is completely known, however, it is also of interest to have some very simple way of describing the probabilistic characteristics. For this purpose the so-called *moments* are introduced here.

When X is a random variable, $Y = X^k$, where k is a positive integer, is also a random variable because $P(\{\omega : X^k \leq y\})$ exists for every y. In the following it is assumed that all random variables are continuous random variables, if not otherwise stated. The *expected value* of X is defined as

$$E[X] = \int_{-\infty}^{\infty} x\, f_X(x)dx \qquad (2.34)$$

The expected value is also called the ensemble average, mean or the *first moment* of X and the symbol μ_X is often used for it. By analogy with this the n'th *moment* of X is called $E[X^n]$ and is defined as

$$E[X^n] = \int_{-\infty}^{\infty} x^n\, f_X(x)dx \qquad (2.35)$$

For discrete random variables the integrals in (2.34) and (2.35) must be replaced by summations.

Note that the first moment of X defined by equation (2.34) is analogous to the location of the centroid of a unit mass. Likewise, the second moment can be compared with the mass moment of inertia.

: **Example 2.9.** Consider the discrete random variable X defined in example 2.7. The
: discrete version of (2.34) gives
:
: $E[X] = 4\cdot 0.3 + 5\cdot 0.5 + 6\cdot 0.2 = 4.9$
:
: The most probable value is called the *mode* and is in this case equal to 5.0 (see figure
: 2.5). Further
:
: $E[X^2] = 16\cdot 0.3 + 25\cdot 0.5 + 36\cdot 0.2 = 24.5$

Above, a new random variable $Y = X^k$ was considered. This is a special case of a random variable which is a function of another random variable whose distribution function is known. Let $Y = f(X)$, where f is a function with at most a finite number of discontinuities. Then it is possible to show that Y is a random variable according to the definition of a random variable. If the function f is monotonic the distribution function F_Y is given by

$$F_Y(y) = P(Y \leqslant y) = P(X \leqslant f^{-1}(y)) = F_X(f^{-1}(y)) \qquad (2.36)$$

and the density function f_Y by

$$f_Y(y) = \frac{d}{dy} F_Y(y) = \frac{d}{dy} F_X(f^{-1}(y)) = f_X(f^{-1}(y)) \cdot \left|\frac{df^{-1}(y)}{dy}\right| \qquad (2.37)$$

or simply

$$f_Y(y) = f_X(x) \left|\frac{dx}{dy}\right| \qquad (2.38)$$

Example 2.10. Let $Y = aX + b$. Then $X = (Y - b)/a$ and

$$f_Y(y) = f_X\left(\frac{y-b}{a}\right) \cdot \left|\frac{1}{a}\right|$$

It is important to note that the expected value of $Y = f(X)$ can be computed in the following way without determining f_Y

$$E[Y] = \int_{-\infty}^{\infty} y\, f_Y(y)dy = \int_{-\infty}^{\infty} f(x)\, f_X(x)dx = E[f(X)] \tag{2.39}$$

Exercise 2.5. Show that

$$E[\sum_{i=1}^{n} f_i(X)] = \sum_{i=1}^{n} E[f_i(X)] \tag{2.40}$$

so that the operations of expectation and summation can commutate.

Returning to the moments of a random variable X, the nth *central moment* of X is defined by $E[(X - \mu_X)^n]$, where $\mu_X = E[X]$. Note that the first central moment of X is always equal to zero. The second central moment of X is called the *variance* of X and is denoted by σ_X^2 or Var[X]

$$\text{Var}[X] = \sigma_X^2 = E[(X - \mu_X)^2] \tag{2.41}$$

The positive square root of the variance, σ_X, is called the *standard deviation* of X.

Exercise 2.6. Show that

$$\sigma_X^2 = E[X^2] - \mu_X^2 \tag{2.42}$$

The standard deviation σ_X is a measure of how closely the values of the random variable X are concentrated around the expected value E[X]. It is difficult only by knowledge of σ_X to decide whether the dispersion should be considered small or large because this will depend on the expected value. However, the *coefficient of variation* V_X, defined by

$$V_X = \frac{\sigma_X}{\mu_X} \tag{2.43}$$

gives better information regarding the dispersion.

: **Example 2.11.** Consider the same discrete random variable X as in example 2.9, where
: $E[X] = 4.9$ and $E[X^2] = 24.5$. The variance, therefore, is
:
: $$\mathrm{Var}[X] = 24.5 - 4.9^2 = 0.49$$
:
: and the standard deviation is
:
: $$\sigma_X = \sqrt{0.49} = 0.7$$
:
: Thus the coefficient of variation is
:
: $$V_X = \frac{0.7}{4.9} = 0.14$$

The third central moment is a measure of the *asymmetry* or *skewness* of the distribution of a random variable. For a continuous random variable it is defined by

$$E[(X-\mu_X)^3] = \int_{-\infty}^{\infty} (x-\mu_X)^3 f_X(x)dx \tag{2.44}$$

2.6 UNIVARIATE DISTRIBUTIONS

In this section some of most widely used probability distributions are introduced. Perhaps the most important distribution is the *normal distribution* also called the *Gaussian distribution*. It is a two-parameter distribution defined by the density function

$$f_X(x) = \frac{1}{\sigma\sqrt{2\pi}} \exp[-\frac{1}{2}(\frac{x-\mu}{\sigma})^2] \tag{2.45}$$

where μ and σ are parameters equal to μ_X and σ_X. This normal distribution will be denoted $N(\mu, \sigma)$.

The distribution function corresponding to (2.45) is given by

$$F_X(x) = \int_{-\infty}^{x} \frac{1}{\sigma\sqrt{2\pi}} \exp[-\frac{1}{2}(\frac{t-\mu}{\sigma})^2]dt \tag{2.46}$$

This integral cannot be evaluated on a closed form. By the substitution

$$s = \frac{t-\mu}{\sigma} \,, \quad dt = \sigma\, ds \tag{2.47}$$

the equation (2.46) becomes

$$F_X(x) = \int_{-\infty}^{\frac{x-\mu}{\sigma}} \frac{1}{\sqrt{2\pi}} \exp(-\frac{1}{2}s^2)ds = \Phi_X(\frac{x-\mu}{\sigma}) \tag{2.48}$$

where Φ_X is the *standard normal distribution function* defined by

$$\Phi_X(x) = \int_{-\infty}^{x} \frac{1}{\sqrt{2\pi}} \exp(-\frac{t^2}{2})dt \tag{2.49}$$

The corresponding *standard normal density function* is

$$\varphi_X(x) = \frac{1}{\sqrt{2\pi}} \exp(-\frac{x^2}{2}) \tag{2.50}$$

Due to the important relation (2.48) only a standard normal table is necessary. The functions φ_X and Φ_X are shown in figure 2.8.

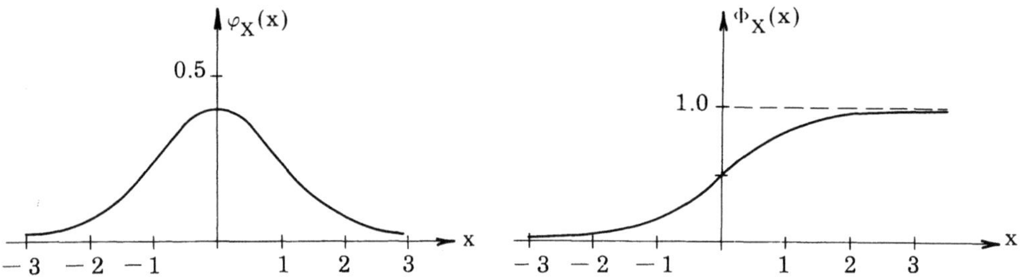

Figure 2.8.

Let the random variable $Y = \ln X$ be normally distributed $N(\mu_Y, \sigma_Y)$. Then the random variable X is said to follow a *logarithmic normal distribution* with the parameters $\mu_Y \in R$ and $\sigma_Y > 0$. The log-normal density function is

$$f_X(x) = \frac{1}{\sigma_Y \sqrt{2\pi}} \frac{1}{x} \exp[-\frac{1}{2}(\frac{\ln x - \mu_Y}{\sigma_Y})^2] \tag{2.51}$$

where $x \geqslant 0$.

: **Exercise 2.7.** Derive the log-normal density function (2.51) by the use of equation (2.38).

Let X be log-normally distributed with the parameters μ_Y and σ_Y. Note that μ_Y and σ_Y are not equal to μ_X and σ_X. It can be shown that

$$\mu_X = \exp(\mu_Y + \frac{1}{2}\sigma_Y^2) \tag{2.52}$$

$$\sigma_X = \sqrt{\mu_X^2 (e^{\sigma_Y^2} - 1)} \tag{2.53}$$

: **Exercise 2.8.** Let X be log-normally distributed with the parameters μ_Y and σ_Y. Show
: that

$$F_X(x) = P(X \leqslant x) = \Phi(\frac{\ln x - \mu_Y}{\sigma_Y}) \tag{2.54}$$

: where Φ is the standard normal distribution function.

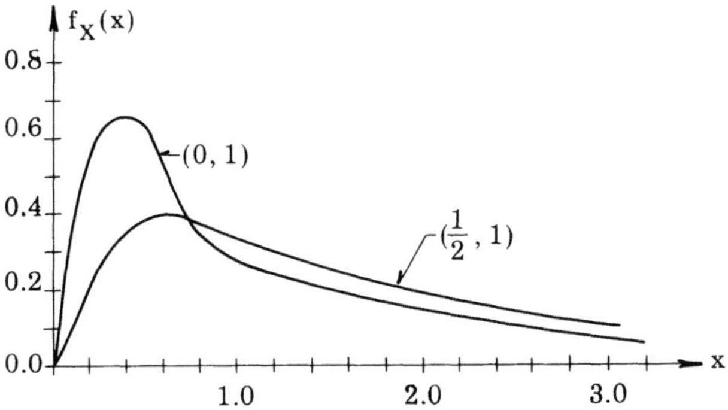

Figure 2.9

The log-normal density functions with the parameters $(\mu_Y, \sigma_Y) = (0, 1)$ and $(1/2, 1)$ are illustrated in figure 2.9.

Example 2.12. Let the compressive strength X for concrete be log-normally distributed with the parameters $(\mu_Y, \sigma_Y) = (3 \text{ MPa}, 0.2 \text{ MPa})$. Then

$$\mu_X = \exp(3 + \frac{1}{2} \cdot 0.04) = 20.49 \text{ MPa}$$

$$\sigma_X^2 = 20.49^2 (1.0408 - 1) = 17.14 \text{ (MPa)}^2$$

$$\sigma_X = 4.14 \text{ MPa}$$

and

$$P(X \leqslant 10 \text{ MPa}) = \Phi((\ell n 10 - 3)/0.2) = \Phi(-3.487) = 2.4 \cdot 10^{-4}$$

An important distribution is the so-called *Weibull distribution* with 3 parameters β, ϵ and k. The density function f_X is defined by

$$f_X(x) = \frac{\beta}{k - \epsilon} (\frac{x - \epsilon}{k - \epsilon})^{\beta - 1} \exp[-(\frac{x - \epsilon}{k - \epsilon})^{\beta}] \tag{2.55}$$

where $x \geqslant \epsilon$ and $\beta > 1$, $k > \epsilon$.

If $\epsilon = 0$ equation (2.55) is

$$f_X(x) = \frac{\beta}{k} (\frac{x}{k})^{\beta - 1} \exp[-(\frac{x}{k})^{\beta}], \ x \geqslant 0 \tag{2.56}$$

The density function (2.56) is called a two-parameter Weibull density function and is shown in figure 2.10. If $\epsilon = 0$ and $\beta = 2$ in (2.55) the density function is identical with the so-called *Rayleigh* density function

$$f_X(x) = \frac{2x}{k^2} \exp[-(\frac{x}{k})^2] \tag{2.57}$$

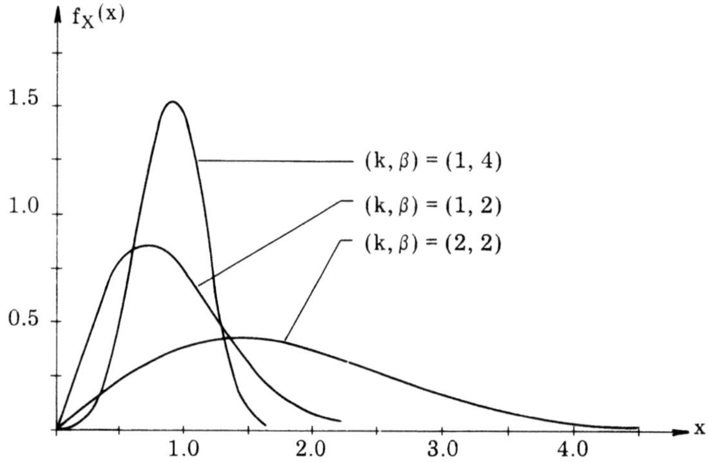

Figure 2.10

2.7 RANDOM VECTORS

Until now, the concept of a random variable has been used only in a one-dimensional sense. In section 2.4 a random variable is defined as a real-valued function $X:\Omega \to R$ mapping the sample space Ω into the real line R. This definition can easily be extended to a vector-valued random variable $\overline{X}:\Omega \to R^n$ called a *random vector (random n-tuple),* where $R^n = R \times R \times \ldots \times R$. An n-dimensional random vector $\overline{X}:\Omega \to R^n$ can be considered an ordered set $\overline{X} = (X_1, X_2, \ldots, X_n)$ of one-dimensional random variables $X_i:\Omega \to R$, $i = 1, \ldots, n$. Note that $X_1, X_2, \ldots, X_n$ are defined on the same sample space Ω.

Let X_1 and X_2 be two random variables. The range of the random vector $\overline{X} = (X_1, X_2)$ is then a subset of R^2 as shown in figure 2.11. Likewise, the range of an n-dimensional random vector is a subset of R^n.

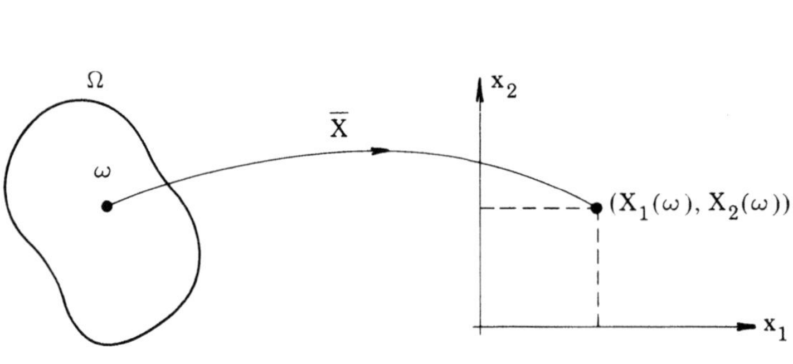

Figure 2.11

Consider again two random variables X_1 and X_2 and the corresponding distribution functions F_{X_1} and F_{X_2}. It is clear that the latter give no information regarding the joint behaviour of X_1 and X_2. To describe the joint behaviour of X_1 and X_2 the *joint probability distribution function* $F_{X_1,X_2}: R^2 \rightarrow R$ is introduced and defined by

$$F_{X_1,X_2}(x_1, x_2) = P((X_1 \leq x_1) \cap (X_2 \leq x_2)) \tag{2.58}$$

It is often convenient to use the notation $F_{\overline{X}}$ for F_{X_1,X_2}, where $\overline{X} = (X_1, X_2)$. The definition (2.58) can be generalized to the n-dimensional case

$$F_{\overline{X}}(\overline{x}) = P(\bigcap_{i=1}^{n} (X_i \leq x_i)) \tag{2.59}$$

where $\overline{X} = (X_1, \ldots, X_n)$ and $\overline{x} = (x_1, \ldots, x_n)$.

A random vector can be discrete or continuous, but the presentation here will be confined to continuous distributions. Only two-dimensional random vectors will be treated because generalization to n-dimensional random vectors is straightforward.

The *joint probability density function* for the random vector $\overline{X} = (X_1, X_2)$ is defined as

$$f_{\overline{X}}(\overline{x}) = \frac{\partial^2}{\partial x_1 \partial x_2} F_{\overline{X}}(\overline{x}) \tag{2.60}$$

The inverse of (2.60) is

$$F_{\overline{X}}(\overline{x}) = \int_{-\infty}^{x_1} \int_{-\infty}^{x_2} f_{\overline{X}}(x_1', x_2') dx_1' dx_2' \tag{2.61}$$

The distribution functions F_{X_1} and F_{X_2} for the single random variables X_1 and X_2 can be obtained from (2.61)

$$F_{X_1}(x_1) = P(X_1 \leq x_1) = \int_{-\infty}^{x_1} \int_{-\infty}^{\infty} f_{\overline{X}}(x_1', x_2') dx_1' dx_2' \tag{2.62}$$

and similarly for F_{X_2}. By differentiation of (2.62)

$$f_{X_1}(x_1) = \int_{-\infty}^{\infty} f_{\overline{X}}(x_1, x_2) dx_2 \tag{2.63}$$

and correspondingly for f_{X_2}

$$f_{X_2}(x_2) = \int_{-\infty}^{\infty} f_{\overline{X}}(x_1, x_2) dx_1 \tag{2.64}$$

The density functions f_{X_1} and f_{X_2} are called *marginal density functions.*

Example 2.13. Consider again example 2.1 and let a 2-dimensional discrete random vector $\bar{X} = (X_1, X_2)$ be defined on Ω by

$$P(4, 3) = 0.1$$
$$P(4, 4) = 0.1$$
$$P(5, 3) = 0.3$$
$$P(5, 4) = 0.2$$
$$P(6, 3) = 0.2$$
$$P(6, 4) = 0.1$$

The mass function $p_{\bar{X}}$ is illustrated in figure 2.12, and the marginal mass functions p_{X_1} and p_{X_2} in figure 2.13.

Note that $p_{\bar{X}}(x_1, x_2) \neq p_{X_1}(x_1) \cdot p_{X_2}(x_2)$.

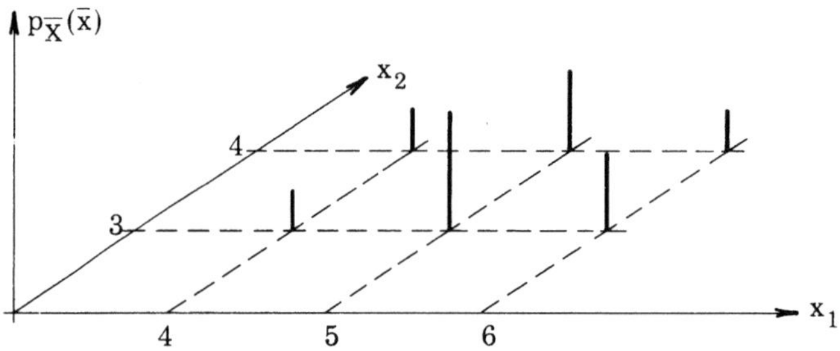

Figure 2.12

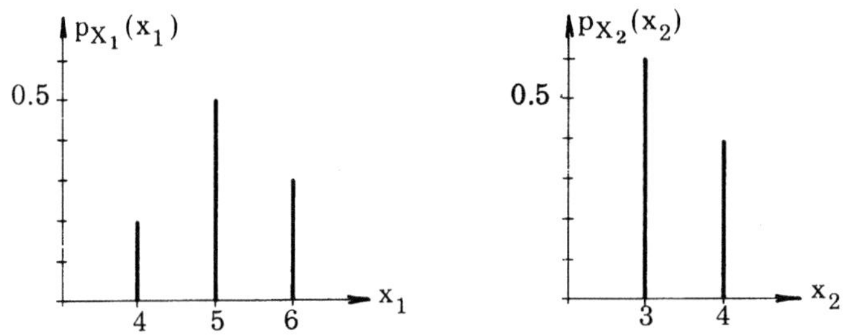

Figure 2.13

2.8 CONDITIONAL DISTRIBUTIONS

In equation (2.16) the probability of occurrence of event E_1 conditional upon the occurrence of event E_2 was defined by

$$P(E_1 \mid E_2) = \frac{P(E_1 \cap E_2)}{P(E_2)} \tag{2.16}$$

In accordance with this definition the *conditional probability mass function* for two jointly distributed discrete random variables X_1 and X_2 is defined as

$$p_{X_1 \mid X_2}(x_1 \mid x_2) = \frac{p_{X_1, X_2}(x_1, x_2)}{p_{X_2}(x_2)} \tag{2.65}$$

A natural extension to the continuous case is the following definition of the *conditional probability density function*

$$f_{X_1 \mid X_2}(x_1 \mid x_2) = \frac{f_{X_1, X_2}(x_1, x_2)}{f_{X_2}(x_2)} \tag{2.66}$$

where $f_{X_2}(x_2) > 0$ and where f_{X_2} is defined by (2.64). Note that $p_{X_1 \mid X_2}$ is a mass function in (2.65) and $f_{X_1 \mid X_2}$ a density function in (2.66).

The two random variables X_1 and X_2 are said to be *independent* if

$$f_{X_1 \mid X_2}(x_1 \mid x_2) = f_{X_1}(x_1) \tag{2.67}$$

which implies

$$f_{X_1, X_2}(x_1 \mid x_2) = f_{X_1}(x_1)\, f_{X_2}(x_2) \tag{2.68}$$

By integrating (2.66) with respect to x_1 one gets the *conditional distribution function*

$$F_{X_1 \mid X_2}(x_1 \mid x_2) = \frac{\int_{-\infty}^{x_1} f_{X_1, X_2}(x_1', x_2)dx_1'}{f_{X_2}(x_2)} \tag{2.69}$$

Next by integrating with respect to x_2 the so-called *total probability theorem* is shown

$$F_{X_1}(x_1) = \int_{-\infty}^{\infty} F_{X_1 \mid X_2}(x_1 \mid x_2)\, f_{X_2}(x_2)dx_2 \tag{2.70}$$

Example 2.14. Consider again the two jointly distributed discrete random variables X_1 and X_2 from example 2.13. Note that

$$p_{X_1, X_2}(5, 3) = p_{X_1}(5) p_{X_2}(3)$$

but for example

$$p_{X_1,X_2}(6, 4) \neq p_{X_1}(6) p_{X_2}(4)$$

Therefore, X_1 and X_2 are not independent.

Exercise 2.9. Consider two jointly distributed discrete random variables X_1 and X_2 with the probability mass functions p_{X_1} and p_{X_2} given in figure 2.13 and assume that X_1 and X_2 are independent. Determine the joint probability mass function $p_{\overline{X}}$ for the random vector $\overline{X} = (X_1, X_2)$.

2.9 FUNCTIONS OF RANDOM VARIABLES

In chapter 2.5 a random variable Y, which is a function f(X) of another random variable X, was treated and it was shown how the density function f_Y could be determined on the basis of the density function f_X, namely by equation (2.38)

$$f_Y(y) = f_X(x) \left| \frac{dx}{dy} \right| \tag{2.38}$$

where $x = f^{-1}(y)$. This will now be generalized to random vectors, where the random vector $\overline{Y} = (Y_1, Y_2, \ldots, Y_n)$ is a function $\overline{f} = (f_1, \ldots, f_n)$ of the random vector $\overline{X} = (X_1, X_2, \ldots, X_n)$, that is

$$Y_i = f_i(X_1, \ldots, X_n) \tag{2.71}$$

where $i = 1, 2, \ldots, n$. It is assumed that the functions f_i, $i = 1, 2, \ldots, n$ are one-to-one functions so that inverse relations exist

$$X_i = g_i(Y_1, \ldots, Y_n) \tag{2.72}$$

It can then be shown that

$$f_{\overline{Y}}(\overline{y}) = f_{\overline{X}}(\overline{x}) \, |J| \tag{2.73}$$

where $\overline{x} = (x_1, \ldots, x_n)$ and $\overline{y} = (y_1, \ldots, y_n)$ and

$$J = \begin{vmatrix} \frac{\partial x_1}{\partial y_1} & \cdots\cdots & \frac{\partial x_1}{\partial y_n} \\ \vdots & & \vdots \\ \frac{\partial x_n}{\partial y_1} & \cdots\cdots & \frac{\partial x_n}{\partial y_n} \end{vmatrix} \tag{2.74}$$

is the Jacobian determinant.

Let the random variable Y be a function f of the random vector $\overline{X} = (X_1, \ldots, X_n)$, that is

$$Y = f(\overline{X}) = f(X_1, \ldots, X_n) \tag{2.75}$$

It can be shown that

$$E[Y] = \int_{-\infty}^{\infty} \cdots \int_{-\infty}^{\infty} f(\bar{x})\, f_{\bar{X}}(\bar{x})dx_1 \ \ldots \ dx_n \tag{2.76}$$

where $\bar{x} = (x_1, \ldots, x_n)$ and $f_{\bar{X}}(\bar{x})$ is the probability density function for the random vector $\bar{X}$.

Exercise 2.10. Show that

$$E[\sum_{i=1}^{n} f_i(X_i)] = \sum_{i=1}^{n} E[f_i(X_i)] \tag{2.77}$$

so that the operation of expectations and summations can commutate. Compare with exercise 2.5.

Exercise 2.11. Show that

$$E[\prod_{i=1}^{n} f_i(X_i)] = \prod_{i=1}^{n} E[f_i(X_i)] \tag{2.78}$$

when $X_1, \ldots, X_n$ are independent random variables.

Let X_1 and X_2 be two random variables with the expected values $E[X_1] = \mu_{X_1}$ and $E[X_2] = \mu_{X_2}$. The *mixed central moment* defined by

$$\mathrm{Cov}[X_1, X_2] = E[(X_1 - \mu_{X_1})(X_2 - \mu_{X_2})] \tag{2.79}$$

is called the *covariance* of X_1 and X_2. The ratio,

$$\rho_{X_1 X_2} = \frac{\mathrm{Cov}[X_1, X_2]}{\sigma_{X_1}\sigma_{X_2}} \tag{2.80}$$

where σ_{X_1} and σ_{X_2} are the standard deviations of the random variables X_1 and X_2, is called the *correlation coefficient*. It can be used as a measure of mutual linear dependence between a pair of random variables. It can be shown that $-1 \leqslant \rho_{X_1 X_2} \leqslant 1$.

Two random variables X_1 and X_2 are said to be *uncorrelated* if $\rho_{X_1 X_2} = 0$. It follows from equation (2.76) that

$$\mathrm{Cov}[X_1, X_2] = E[(X_1 - \mu_{X_1})(X_2 - \mu_{X_2})] = E[X_1 \cdot X_2] - E[X_1]\ E[X_2] \tag{2.81}$$

Therefore, for uncorrelated random variables X_1 and X_2 we have

$$E[X_1 \cdot X_2] = E[X_1]E[X_2] \tag{2.82}$$

It is important to note that independent random variables are uncorrelated, but uncorrelated variables are not in general independent.

Note that

$$\mathrm{Cov}[X_i, X_i] = \mathrm{Var}[X_i] \tag{2.83}$$

Therefore, the mutual correlation between random variables $X_1, X_2, \ldots, X_n$ can be expressed by the so-called *covariance matrix* $\bar{\bar{C}}$ defined by

$$\bar{\bar{C}} = \begin{bmatrix} \mathrm{Var}[X_1] & \mathrm{Cov}[X_1, X_2] & \cdots\cdots & \mathrm{Cov}[X_1, X_n] \\ \mathrm{Cov}[X_2, X_1] & \mathrm{Var}[X_2] & \cdots\cdots & \mathrm{Cov}[X_2, X_n] \\ \vdots & \vdots & & \vdots \\ \mathrm{Cov}[X_n, X_1] & \mathrm{Cov}[X_n, X_2] & \cdots\cdots & \mathrm{Var}[X_n] \end{bmatrix} \tag{2.84}$$

Exercise 2.12. Let the random variable Y be defined by

$$Y = a_1 X_1 + a_2 X_2 \tag{2.85}$$

where X_1, X_2 are random variables and a_1, a_2 constants. Show that

$$\mathrm{Var}[Y] = a_1^2 \mathrm{Var}[X_1] + a_2^2 \mathrm{Var}[X_2] + 2a_1 a_2 \mathrm{Cov}[X_1, X_2] \tag{2.86}$$

2.10 MULTIVARIATE DISTRIBUTIONS

The most important joint density function of two continuous random variables X_1 and X_2 is the *bivariate normal density function* given by

$$f_{X_1,X_2}(x_1, x_2) = \frac{1}{2\pi\sigma_1\sigma_2\sqrt{1-\rho^2}} \exp[\frac{-1}{2(1-\rho^2)} ((\frac{x_1-\mu_1}{\sigma_1})^2$$

$$-2\rho(\frac{x_1-\mu_1}{\sigma_1})(\frac{x_2-\mu_2}{\sigma_2}) + (\frac{x_2-\mu_2}{\sigma_2})^2)] \tag{2.87}$$

where $-\infty \leqslant x_1 \leqslant \infty$, $-\infty \leqslant x_2 \leqslant \infty$, and μ_1, μ_2 are the means, σ_1, σ_2 the standard deviations and ρ the correlation coefficient of X_1 and X_2.

Exercise 2.13. Show that the marginal density functions $f_{X_1}(x_1)$ for (2.87) are

$$f_{X_1}(x_1) = \frac{1}{\sqrt{2\pi}\,\sigma_1} \exp[-\frac{1}{2} (\frac{x_1-\mu_1}{\sigma_1})^2] \tag{2.88}$$

The *multivariate normal density function* is defined as

$$f_{\bar{X}}(\bar{x}) = \frac{1}{(2\pi)^{\frac{n}{2}}} \cdot \frac{1}{|\bar{\bar{C}}|^{\frac{1}{2}}} \exp[-\frac{1}{2} \sum_{i,j=1}^{n} (x_i - \mu_i) M_{ij} (x_j - \mu_i)] \qquad (2.89)$$

where $\bar{x} = (x_1, x_2, \ldots, x_n)$, $\bar{\bar{M}} = \bar{\bar{C}}^{-1}$, and where $\bar{\bar{C}}$ is the covariance matrix defined by (2.84).

BIBLIOGRAPHY

[2.1] Ang, A. H-S. & W. H. Tang: *Probability Concepts in Engineering Planning and Design.* Vol. 1, Wiley, N. Y., 1975.

[2.2] Benjamin, J. R. & C. A. Cornell: *Probability, Statistics and Decision for Civil Engineers.* McGraw-Hill, N.Y., 1970.

[2.3] Bolotin, V. V.: *Statistical Methods in Structural Mechanics.* Holden-Day, San Francisco, 1969.

[2.4] Ditlevsen, O.: *Uncertainty Modeling.* McGraw-Hill, N.Y., 1981.

[2.5] Lin, Y. K.: *Probabilistic Theory of Structural Dynamics.* McGraw-Hill, N.Y., 1967.

[2.6] Feller, W.: *An Introduction to Probability Theory and its Applications.* Wiley, N.Y., Vol. I, 1950, Vol. II, 1966.

[2.7] Larson, H. J. & B. O. Shubert: *Probabilistic Models in Engineering Sciences.* T. Wiley & Sons, N.Y., Vol. I & II, 1979.

Chapter 3

PROBABILISTIC MODELS FOR LOADS AND RESISTANCE VARIABLES

3.1 INTRODUCTION

In this chapter the aim is to examine the way in which suitable probabilistic models can be developed to represent the uncertainties that exist in typical basic variables. We shall first consider the problem of modelling physical variability and then turn to the question of incorporating statistical uncertainty.

Load and resistance parameters clearly require different treatment, since loads are generally time-varying. As discussed in chapters 9 and 10, time-varying loads are best modelled as stochastic processes, but this is not a convenient representation for use with the methods of reliability analysis being presented here (chapters 5 and 6). Instead, it is appropriate to use the distribution of the extreme value of the load in the reference period for which the reliability is required; or, where there are two or more time-varying loads acting on a structure together, the distribution of the extreme combined load or load effect. The particular problems associated with the analysis of combined loading are discussed in chapter 10.

The selection of probabilistic models for basic random variables can be divided into two parts - the choice of suitable probability distributions with which to characterize the physical uncertainty in each case and the choice of appropriate values for the parameters of those distributions. For most practical problems neither task is easy since there may be a number of distributions which appear to fit the available data equally well. As mentioned above, loads and resistance variables require different treatment and will be discussed separately. However, it is first necessary to introduce the important subject of the statistical theory of extremes which is of relevance to both load and strength variables. This topic is discussed in the next two sections.

3.2 STATISTICAL THEORY OF EXTREMES

In the modelling of loads and in the reliability analysis of structural systems it is necessary to deal with the theory of extreme values. For example, with time-varying loads, the analyst is interested in the likely value of the greatest load during the life of the structure. To be more precise, he wishes to know the probability distribution of the greatest load. This may be interpreted physically as the distribution that would be obtained if the maximum lifetime load were measured in an infinite set of nominally identical structures.

In an analogous way, if the strength of a structure depends on the strength of the weakest of a number of elements - for example, a statically determinate truss - one is concerned with the probability distribution of the minimum strength.

In general, one can estimate from test results or records the parameters of the distribution of the instantaneous values of load or of the strength of individual components, and from this information the aim is to derive the distribution for the smallest or largest values.

3.2.1 Derivation of the cumulative distribution of the ith smallest value of n identically distributed independent random variables X_i

Assume the existence of a random variable X (e.g. the maximum mean-hourly wind speed in consecutive yearly periods) having a cumulative distribution function F_X and a corresponding probability density function f_X. This is often referred to as the *parent distribution*. Taking a sample size of n (e.g. n years records and n values of the maximum mean-hourly wind speed) let the cumulative distribution function of the ith smallest value X_i^n in the sample be $F_{X_i^n}$ and its corresponding density function be $f_{X_i^n}$.

Then

$$\begin{aligned} f_{X_i^n}(x)dx &= \text{constant} \times \text{probability that } (i-1) \text{ values of X fall below x} \\ &\quad \times \text{probability that } (n-i) \text{ values of X fall above x} \\ &\quad \times \text{probability that 1 value of X lies in the range x to } (x+dx) \\ &= c F_X^{i-1}(x)(1-F_X(x))^{n-i} f_X(x)dx \end{aligned} \tag{3.1}$$

where

$$c = \frac{n(n-1)!}{(i-1)!(n-i)!} = \text{the number of ways of choosing } (i-1) \text{ values less than x, together with } (n-i) \text{ values greater than x} \tag{3.2}$$

Thus

$$F_{X_i^n}(y) = \int_0^y f_{X_i^n}(x)dx = \int_0^y c F_X^{i-1}(x)(1-F_X(x))^{n-i} f_X(x)dx \tag{3.3}$$

This can be shown to be equal to

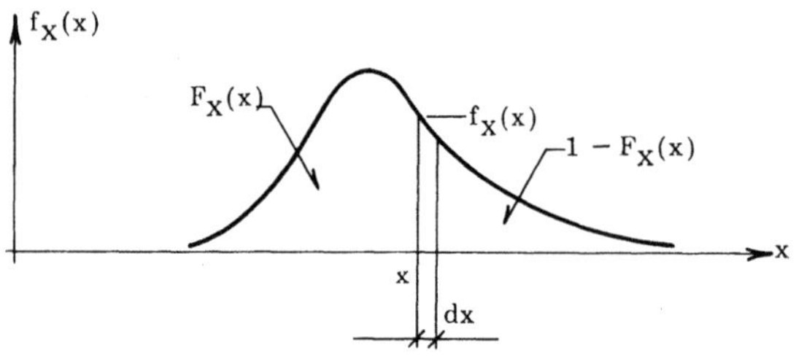

Figure 3.1

$$F_{X_i^n}(y) = \frac{n(n-1)!}{(i-1)!(n-i)!}\left[\frac{(F_X(y))^i}{i} - \binom{n-i}{1}\frac{(F_X(y))^{i+1}}{(i+1)} + \binom{n-i}{2}\times \frac{(F_X(y))^{i+2}}{(i+2)} \ldots + (-1)^{(n-i)}\binom{n-i}{n-i}\frac{(F_X(y))^n}{n}\right] \tag{3.4}$$

Exercise 3.1. Show that equation (3.4) can be derived from equation (3.3) by expanding $(1-F_X(x))^{n-i}$ and integrating by parts.

Equation (3.4) gives the probability distribution function for the ith smallest value of n values sampled at random from a variable X with a probability distribution F_X.

Two special cases will now be considered in the following examples.

Example 3.1. For i = n equation (3.4) simplifies to:

$$F_{X_n^n}(x) = (F_X(x))^n \tag{3.5}$$

This is the distribution function for the maximum value in a sample size n.

Example 3.2. For i = 1 equation (3.4) simplifies to:

$$F_{X_1^n}(x) = 1-(1-F_X(x))^n \tag{3.6}$$

This is the distribution function for the minimum value in a sample size n.

It should be noted that $F_{X_n^n}(x)$ may also be interpreted as the probability of the non-occurrence of the event $(X > x)$ in any of n independent trials, so that equation (3.5) follows immediately from the multiplication rule for probabilities. Equation (3.6) may be interpreted in an analogous manner. See also chapter 7.

3.2.2 Normal extremes

If a random variable is normally distributed with mean μ_X with standard deviation σ_X the variable has a distribution function F_X (see (2.46))

$$F_X(x) = \int_{-\infty}^{x} \frac{1}{\sqrt{2\pi}}\,\frac{1}{\sigma_X}\exp\left(-\frac{1}{2}\left(\frac{t-\mu_X}{\sigma_X}\right)^2\right)dt \tag{3.7}$$

If we are interested in the distribution of the maximum value of n identically distributed normal random variables with parameters μ_X and σ_X this has a distribution function

$$F_{X_n^n}(x) = \left(\int_{-\infty}^{x} \frac{1}{\sqrt{2\pi}}\,\frac{1}{\sigma_X}\exp\left(-\frac{1}{2}\left(\frac{t-\mu_X}{\sigma_X}\right)^2\right)dt\right)^n \tag{3.8}$$

It should be noted that $F_{X_n^n}$ is *not normally distributed.*

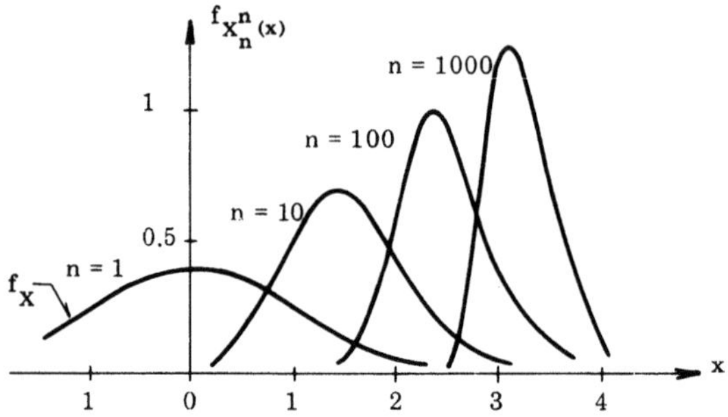

Figure 3.2.

The probability density function $f_{X_n^n} = \frac{d}{dx}(F_{X_n^n})$ is shown in figure 3.2 for various values of n and with X distributed N(0, 1).

3.3 ASYMPTOTIC EXTREME-VALUE DISTRIBUTIONS

It is fortunate that for a very wide class of parent distributions, the distribution functions of the maximum or minimum values of large random samples taken from the parent distribution tend towards certain limiting distributions as the sample becomes large. These are called *asymptotic extreme-value distributions* and are of three main types, I, II and III.

For example, if the particular variable of interest is the maximum of many similar but independent events (e.g. the annual maximum mean-hourly wind speed at a particular site) there are generally good theoretical grounds for expecting the variable to have a distribution function which is very close to one of the asymptotic extreme value distributions. For detailed information on this subject the reader should refer to a specialist text, e.g. Gumbel [3.8] or Mann, Schafer and Singpurwalla [3.11]. Only the most frequently used extreme-value distributions will be referred to here.

3.3.1 Type I extreme-value distributions (Gumbel distributions)

Type I asymptotic distribution of the largest extreme: If the upper tail of the parent distribution falls off in an exponential manner, i.e.

$$F_X(x) = 1 - e^{-g(x)} \tag{3.9}$$

where g is an increasing function of x, then the distribution function F_Y of the largest value Y, from a large sample selected at random from the parent population, will be of the form

$$F_Y(y) = \exp(-\exp(-\alpha(y-u))) \quad , \quad -\infty \leqslant y \leqslant \infty \quad , \quad \alpha > 0 \tag{3.10}$$

Formally, F_Y will asymptotically approach the distribution given by the right hand side of equation (3.10) as $n \to \infty$.

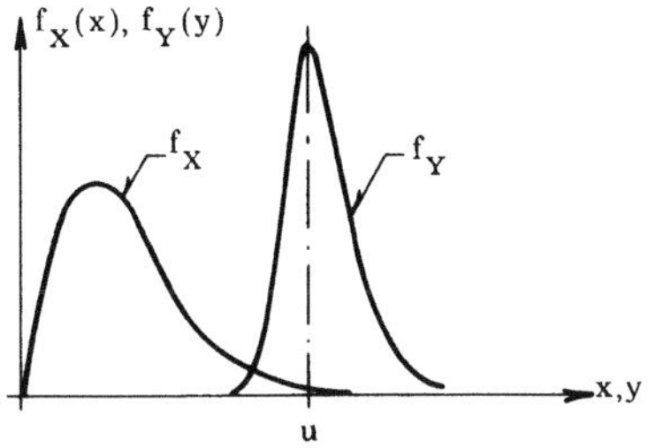

Figure 3.3

The parameters u and α are respectively measures of location and dispersion. u is the *mode* of the asymptotic extreme-value distribution (see figure 3.3).

The mean and standard deviation of the type I maxima distribution (3.10) are related to the parameters u and α as follows

$$\mu_Y = u + \frac{\gamma}{\alpha} \simeq u + \frac{0.5772}{\alpha} \tag{3.11}$$

and

$$\sigma_Y = \frac{\pi}{\sqrt{6}\,\alpha} \tag{3.12}$$

where γ is Euler's constant. This distribution is positively skew as shown in figure 3.3.

A useful property of the type I maxima distribution is that the distribution function $F_{Y_n^n}$ for the largest extreme in any sample of size n is also type I maxima distributed. Furthermore, the standard deviation remains constant (is independent of n), i.e.

$$\sigma_{Y_n^n} = \sigma_Y \tag{3.13}$$

This property is of help in the analysis of load combinations when different numbers of repetitions of loads n_i need to be considered (see chapter 10). In this connection, it is useful to be able to calculate the parameters of the extreme variable Y_n^n from a knowledge of the parameters of Y.

If Y is type I maxima distributed with distribution function F_Y given by equation (3.10) and with parameters α and u, then the extreme distribution of maxima generated in n independent trials has a distribution function

$$F_{Y_n^n}(y) = \exp(-n\exp(-\alpha(y-u))) \tag{3.14}$$

with mean given by

$$\mu_{Y_n^n} = \mu_Y + \frac{\sqrt{6}}{\pi}\sigma_Y \ln(n) \tag{3.15}$$

Type I asymptotic distribution of the smallest extreme: This is of rather similar form to the Type I maxima distribution, but will not be discussed here. The reader should refer to one of the standard texts - see [3.8], [3.11] or [3.5].

3.3.2 Type II extreme-value distributions

As with the type I extreme-value distributions, the type II distributions are of two types. Only the type II distribution of the largest extreme will be considered here. Its distribution function F_Y is given by

$$F_Y(y) = \exp(-(u/y)^k) \ , \ y \geqslant 0, u > 0, k > 0 \tag{3.16}$$

where the parameters u and k are related to the mean and standard deviation by

$$\mu_Y = u\Gamma(1-1/k) \qquad \text{with} \quad k > 1 \tag{3.17}$$

$$\sigma_Y = u[\Gamma(1-2/k) - \Gamma^2(1-1/k)]^{\frac{1}{2}} \qquad \text{with} \quad k > 2 \tag{3.18}$$

where Γ is the gamma function defined by

$$\Gamma(k) = \int_0^\infty e^{-u} u^{k-1} du \tag{3.19}$$

It should be noted that for $k \leqslant 2$, the standard deviation σ_Y is not defined. It is also of interest that if Y is type II maxima distributed, then $Z = \ln Y$ is type I maxima distributed.

: **Exercise 3.2.** Let Y be type II maxima distributed with distribution function F_Y and
: coefficient of variation σ_Y/μ_Y. Show that the variable representing the largest extreme with
: distribution function $(F_Y(y))^n$ has the same coefficient of variation.

The type II maxima distribution is frequently used in modelling extreme hydrological and meterological events. It arises as the limiting distribution of the largest value of many independent identically distributed random variables, when the parent distribution is limited to values greater than zero and has an infinite tail to the right of the form

$$F_X(x) = 1 - \beta(1/x)^k \tag{3.20}$$

3.3.3 Type III extreme-value distributions

In this case only the type III asymptotic distribution of the smallest extreme will be considered. It arises when the parent distribution is of the form:

$$F_X(x) = c(x - \epsilon)^k \qquad \text{with} \quad x \geqslant \epsilon \tag{3.21}$$

i.e. the parent distribution is limited to the left at a value $x = \epsilon$.

In many practical cases ϵ may be zero (i.e. representing a physical limitation on, say, strength).

The distribution of the minimum Y of n independent and identically distributed variables X_i asymptotically approaches the form

$$F_Y(y) = 1 - \exp(-(\frac{y - \epsilon}{k - \epsilon})^\beta) \qquad \text{with} \quad y \geqslant \epsilon, p > 0, k > \epsilon \geqslant 0 \tag{3.22}$$

as $n \to \infty$.

The mean and standard deviation of Y are:

$$\mu_Y = \epsilon + (k - \epsilon)\Gamma(1 + \frac{1}{\beta}) \tag{3.23}$$

and

$$\sigma_Y = (k - \epsilon)(\Gamma(1 + \frac{2}{\beta}) - \Gamma^2(1 + \frac{1}{\beta}))^{\frac{1}{2}} \tag{3.24}$$

The type III minima distribution (3.22) is often known as the 3-parameter *Weibull* distribution and has frequently been used for the treatment of fatigue and fracture problems.

For the special case $\epsilon = 0$, the distribution simplifies to the so-called 2-parameter Weibull distribution

$$F_Y(y) = 1 - e^{-(\frac{y}{k})^\beta} \tag{3.25}$$

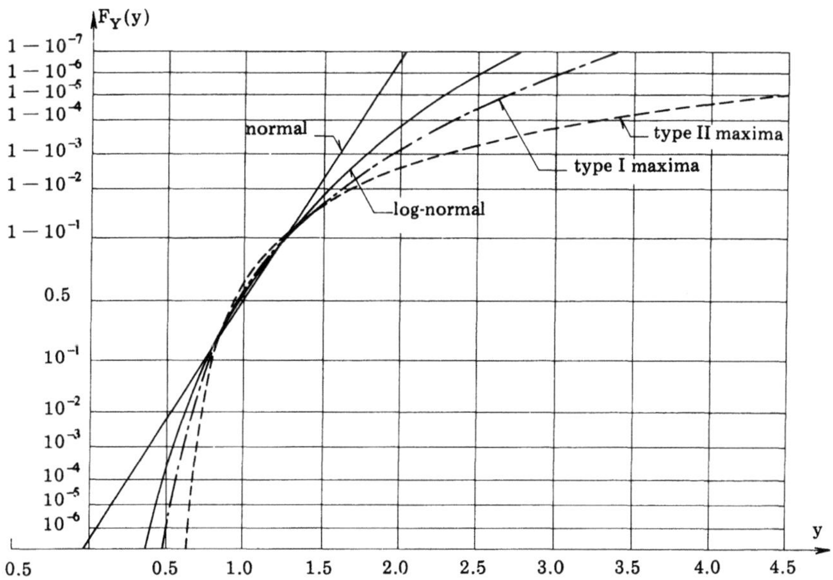

Figure 3.4. Cumulative distributions of different distribution functions ($\mu_Y = 1, \sigma_Y = 0.2$).

with

$$\mu_Y = k\Gamma(1 + \frac{1}{\beta}) \tag{3.26}$$

and

$$\sigma_Y = k(\Gamma(1 + \frac{2}{\beta}) - \Gamma^2(1 + \frac{1}{\beta}))^{\frac{1}{2}} \tag{3.27}$$

Comparisons of the type I maxima and type II maxima distributions with the normal and log-normal distributions are shown in figure 3.4. The random variables in each case have the same mean and standard deviation, namely 1.0 and 0.2.

3.4 MODELLING OF RESISTANCE VARIABLES - MODEL SELECTION

3.4.1 General remarks

In this section some general guidelines are given for the selection of probability distributions to represent the physical uncertainty in variables which affect the strength of structural components and complete structures - for example, dimensions, geometrical imperfections and material properties. Since each material and mechanical property is different, each requires individual attention. Nevertheless, a number of general rules apply. Attention will be restricted here to the modelling of continuously distributed as opposed to discrete quantities.

The easiest starting point is to consider the probability density function f_X of a random variable X as the limiting case of a histogram of sample observations as the number of sample elements is increased and the class interval reduced. However, for small sample sizes, the shape of the histogram varies somewhat from sample to sample, as a result of the random nature of the variable. Figure 3.5 shows two sets of 100 observations of the thickness T of reinforced concrete slabs having a nominal thickness of 150 mm, which illustrates this point. These data were not, in fact, obtained by measurements in real structures but were randomly sampled from a logarithmic normal distribution with a mean μ_T = 150 mm and a coefficient of variation V_T = 0.15 (see appendix A). The corresponding density function f_T is also shown in figure 3.5.

For comparison, figure 3.6 shows data obtained from a real construction site.

A clear distinction must be made, however, between a histogram or a relative frequency diagram on the one hand and a probability density function on the other. Whereas the former is simply a record of observations, the latter is intended for predicting the occurrence of future events - e.g. a thickness less than 100 mm.

If the probability density function f_X of a random variable X is interpreted as the limiting case of a histogram or relative frequency diagram as the sample size tends to infinity, the probability P given by

$$P = P(x_1 < X \leqslant x_2) = \int_{x_1}^{x_2} f_X(x)dx \tag{3.28}$$

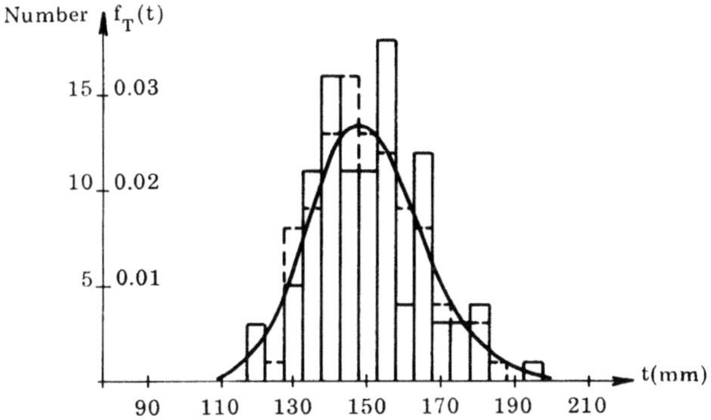

Figure 3.5

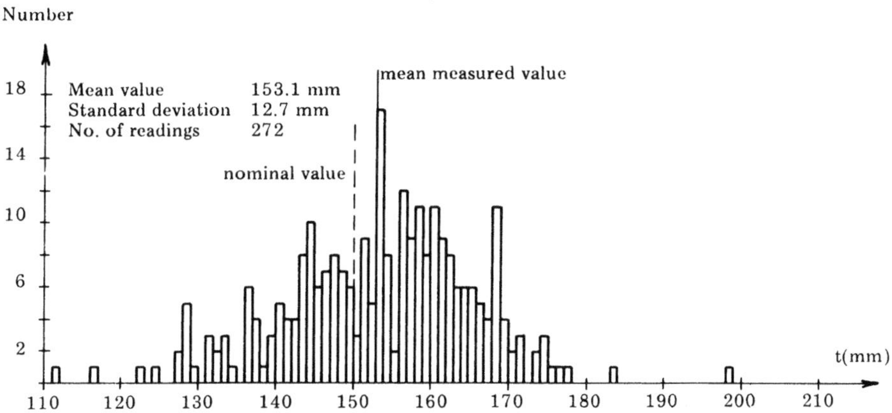

Figure 3.6. Histogram of slab thickness measurements.

clearly has a relative frequency interpretation; i.e. if a very large sample of variable X is obtained at random, the proportion of values within the sample falling in the range $x_1 < X \leqslant x_2$ is likely to to be very close to P. However, this interpretation may not in practice be too helpful. All that can be said is that *if* a variable X does in fact have a known probability density function f_X, and if it is sampled at random an infinite number of times, the proportion in the range $]\, x_1, x_2\, [$ will be P.

The problem of modelling is completely different. In general the engineer is likely to have only a relatively small sample of actual observations of X, along with some prior information obtained from a different source. The problem then is how best to use all this information. Before this question can be answered it is necessary to define exactly what the variable X represents. This is best explained by means of an example.

Example 3.3. Consider the mechanical properties of a single nominal size of continuously-cast hot-rolled reinforcing steel. Let us restrict our attention to a single property, the dynamic yield stress, σ_{yd}, determined at a controlled strain rate of 300 micro-strain per minute and defined as the average height of the stress-strain curve between strains of 0.003 and 0.005, i.e.

$$\sigma_{yd} = \frac{1}{0.002} \int_{\epsilon=0.003}^{\epsilon=0.005} \sigma_y(\epsilon)\, d\epsilon \tag{3.29}$$

where $\sigma_y(\epsilon)$ is the dynamic yield stress at strain ϵ.

Let us assume that this property can be measured with negligible experimental error and that all the reinforcing bars from a single cast of steel are cut into test specimens 0.5 m long and then tested. If σ_{yd} is plotted against Z, the position in the bar, the outcome will be of the form shown in figure 3.7. This is an example of a step-wise continuous-state/continuous-time stochastic process X(t) in which the parameter t may be interpreted as the distance Z along the reinforcing bar. (See chapter 9 for further details of stochastic processes).

The process is interrupted approximately every 600 m because the continuously cast steel is cut into ingots and these are re-heated and rolled separately. The fluctuations in yield stress within each 600 m length are typically very small, i.e. in the order of 1 - 2 N/mm^2. For each 600 m length ℓ, the spatial average yield stress $\bar{\sigma}_{yd}$ is defined as

$$\bar{\sigma}_{yd} = \frac{1}{\ell} \int_0^{\ell} \sigma_{yd}\, d\ell \tag{3.30}$$

The variations in $\bar{\sigma}_{yd}$ from one rolled length to another are typically larger than the within-length variations and are caused mainly by differences in the temperature of the ingot at the start of rolling and by a number of other factors. Some typical data giving values of $\bar{\sigma}_{yd}$ for consecutive lengths of 20 mm diameter hot-rolled high-yield bars from the same cast of steel are shown in figure 3.8 (along with values for the static yield stress). These can be considered as a continuous-state/discrete-time stochastic process. It can be seen that there is a fairly strong positive correlation between $\bar{\sigma}_{yd}$ for adjacent lengths, as might be expected.

If ℓ_c is the total length of reinforcement produced from a single cast of steel then the average yield stress for the cast can be defined as

$$\bar{\bar{\sigma}}_{yd} = \frac{1}{\ell_c} \int_0^{\ell_c} \sigma_{yd}\, d\ell \tag{3.31}$$

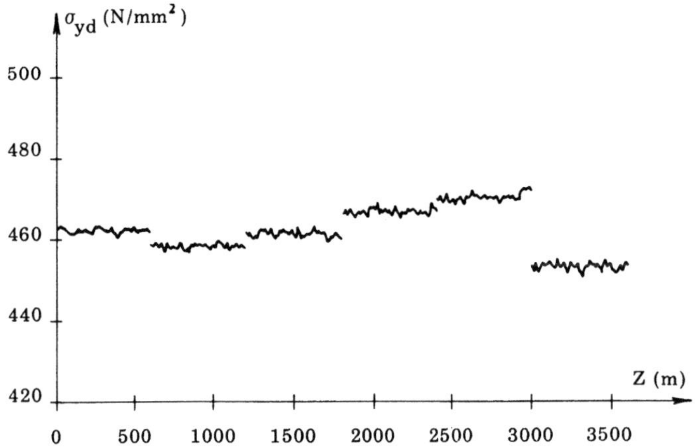

Figure 3.7 Variations in dynamic yield stress along a 20 mm diameter hot-rolled reinforcing bar.

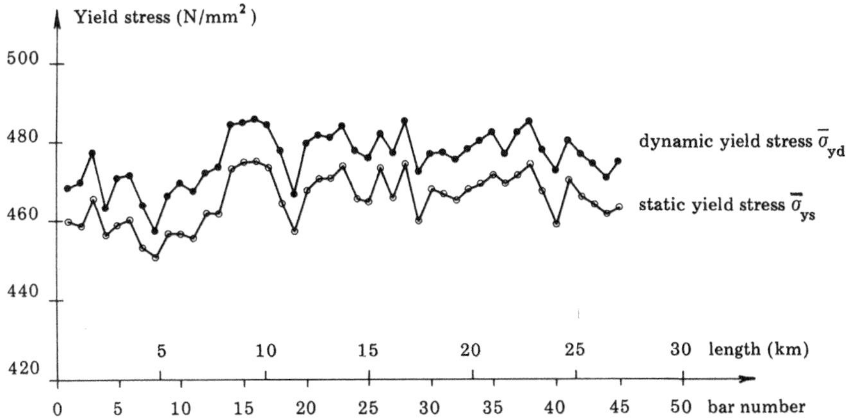

Figure 3.8. Within-cast variations in the yield stress of a 20 mm diameter hot-rolled reinforcing bar.

Provided that the variations in yield stress along each 600 m length of continuously rolled bar can be assumed to be small in comparison with variations in $\bar{\sigma}_{yd}$, the average yield stress for the cast may be expressed as

$$\bar{\bar{\sigma}}_{yd} = \frac{1}{n} \sum_{i=1}^{n} \sigma_{yd}(i) \tag{3.32}$$

where $\sigma_{yd}(i)$ is the yield stress of the ith bar and n is the number of bars rolled from the cast.

If we are interested in the statistical distribution of the yield stress of reinforcing bars supplied to a construction site, account must also be taken of the variations in $\bar{\bar{\sigma}}_{yd}$ that occur from cast to cast. If the steel is to be supplied by a single manufacturer and very close control is exercised over the chemical composition of each cast, variations in $\bar{\bar{\sigma}}_{yd}$ will be very small; but if the chemistry is not well controlled significant differences between casts can occur. If bars are supplied by a number of different manufacturers, systematic differences between manufacturers will be evident even for nominally identical products (e.g. 20 mm diameter bars) because of differences in rolling procedures.

A final effect which must be taken into account is the systematic change in mean yield stress with bar diameter as illustrated in figure 3.9. This phenomenon is quite marked and is rarely taken into account in structural design.

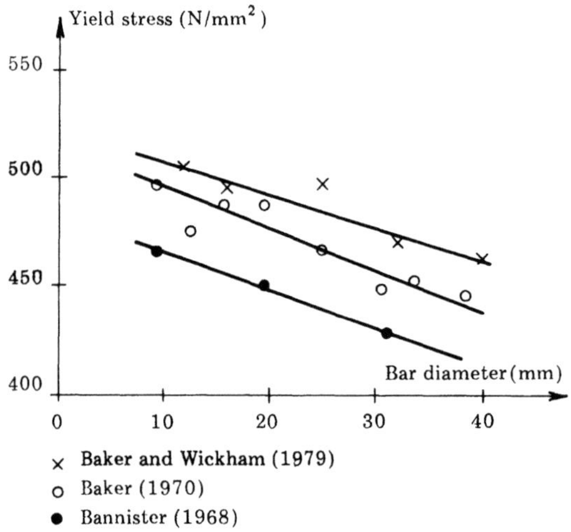

Figure 3.9. Mean yield stress for hot-rolled high yield bars of different diameters.

From the preceding example it is clear that there are many sources of physical variability which contribute to the overall uncertainty in the yield stress of a grade of reinforcing steel. Let us now define the quantity X as the random variable representing the yield stress of a particular grade of reinforcing steel irrespective of source and where »yield stress» is defined in a precise way. We now wish to establish a suitable probability density function for X to use in further calculations. It is clear that the mathematical form of f_x will depend on the particular subset of X, e.g.:

Let A_i be the event [bars are supplied by manufacturer i]

B_j be the event [bars are of diameter j]

C be the event [bars are from a single cast of steel]

Then in general the density functions f_X, $f_{X|A_i}$, $f_{X|B_j}$, $f_{X|A_i \cap B_j}$, $f_{X|A_i \cap B_j \cap C}$ etc. will all be different; not only their parameters but also their shapes. It is also clear that the probability density function f_X representing all bars, irrespective of size or manufacturer, will not be of a simple or standard form (e.g. normal, lognormal, etc.). It will take the form

$$f_X(x) = p_1 f_{X|A_1}(x) + p_2 f_{X|A_2}(x) + \ldots + p_n f_{X|A_n}(x) \quad , \quad \sum_1^n p_i = 1 \tag{3.33}$$

where p_i is the probability that a bar will be supplied by manufacturer i and where

$$f_{X|A_1}(x) = q_1 f_{X|A_1 \cap B_1}(x) + q_2 f_{X|A_1 \cap B_2}(x) + \ldots + q_m f_{X|A_1 \cap B_m}(x) \quad , \quad \sum_1^m q_j = 1 \tag{3.34}$$

q_j being the probability that the bar is of diameter j.

Equation (3.33) represents what is known as a *mixed distribution model.*

It should be noted that because of the systematic decrease in reinforcing bar yield stress with increasing diameter, equation (3.34) gives rise to a density function $f_{X|A_i}$ which is flatter and has less pronounced tails (platykurtic) than any of the component distributions $f_{X|A_i \cap B_j}$. Furthermore, it is generally found that the density function $f_{X|B_j}$ representing bars of a particular size considered over all manufacturers is highly positively skew. The reason for this is discussed in example 3.4 below.

: **Example 3.4.** The yield stress of hot-rolled steel plates of a *single nominal thickness* and
: grade of steel, supplied by a *single* manufacturer, can be shown to be closely represented
: by a log-normal probability distribution (see equation (2.51)), as illustrated by the cumula-
: tive frequency diagrams in figure 3.10. If, however, data from a number of manufacturers
: are combined, the distribution becomes highly skew. This is because manufacturers with
: high product variability have to aim for higher mean properties than manufacturers whose
: products can be closely controlled to achieve the same specified yield stress, for a given
: probability of rejection. See figure 3.11. It should be noted that the scales chosen in figures
: 3.10 and 3.11 are such that a logarithmic normal distribution plots as a straight line.

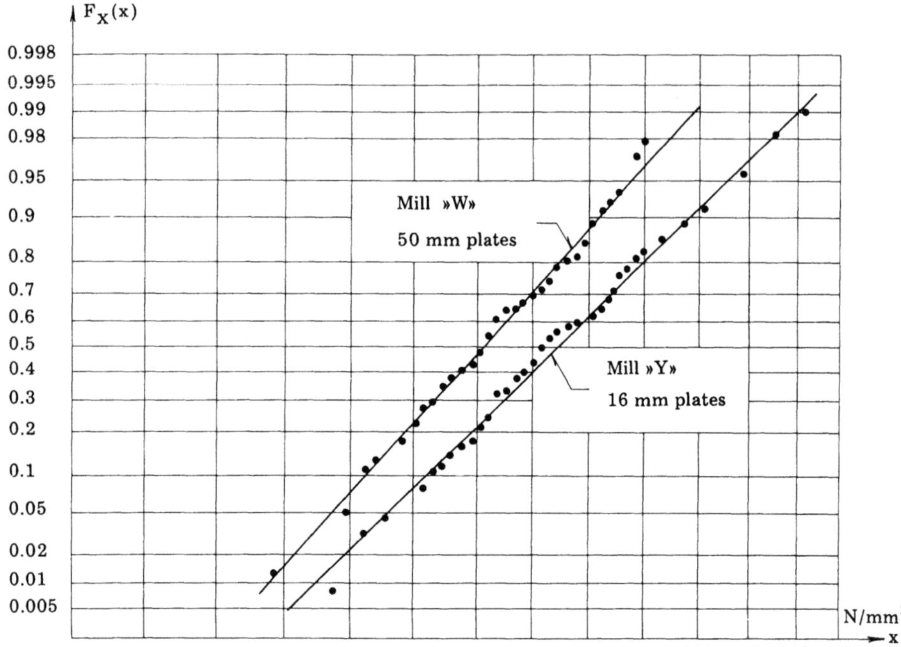

Figure 3.10. Cumulative frequency diagram for yield stress of mild steel plates.

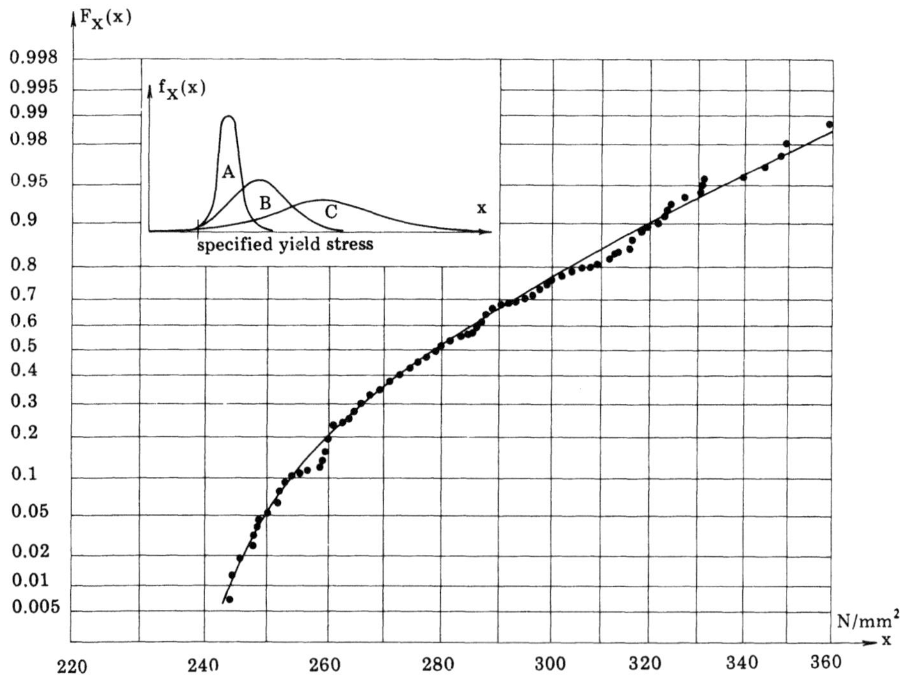

Figure 3.11. Combined cumulative frequency diagram for 12 mm mild steel plates from three mills.

We now return to the question of selecting a suitable probability distribution to model the uncertainty in the strength variable X. It should be clear from the preceding arguments that a procedure of random sampling and testing of, say, reinforcing bars at a construction site and attempts to fit a standard probability distribution to the data will not lead to a sensible outcome. In particular, such a distribution will behave poorly as a predictor of the occurrence of values of X outside the range of the sample obtained. The only sensible approach is to synthesise the probability distribution of X from a knowledge of the component sources of uncertainty (as in equation (3.33)). Admittedly this approach can be adopted only when such information is available. Expressing this problem in another way, it is important that the statistical analysis of data should be restricted to samples which are homogeneous (or more precisely - for which there is no evidence of non-homogeneity).

A further aspect of modelling must now be introduced. Models do not represent reality, they only approximate it. As is well known in other branches of engineering, any one of a number of different empirical models may often be equally satisfactory for some particular purpose, e.g. finite-element versus finite-difference approaches. The same is true of probabilistic models. The question that must be asked is whether the model is suitable for the particular application where it is to be used.

For most structural reliability calculations, the analyst is concerned with obtaining a good fit in the lower tails of the strength distributions, but this may not always be important - for example, when the strength of a structural member is governed by the sum of the strengths of its components. This is illustrated by the following example.

Example 3.5. Consider an axially-loaded reinforced concrete column, a cross-section of which is shown in figure 3.12. If, for the sake of simplicity, the load-carrying capacity of the column is assumed to be given exactly by:

$$R = r_c + \sum_{i=1}^{12} R_i \tag{3.35}$$

where r_c is the load-carrying capacity of the concrete (assumed known) and R_i is the random load-carrying capacity of the ith reinforcing bar at yield. Then, if the various R_i are statistically independent,

$$E[R] = E[r_c + \sum_{i=1}^{12} R_i] = r_c + \sum_{i=1}^{12} E[R_i] \tag{3.36}$$

and

$$Var[R] = Var[r_c + \sum_{i=1}^{12} R_i] = \sum_{i=1}^{12} Var[R_i] \tag{3.37}$$

i.e.

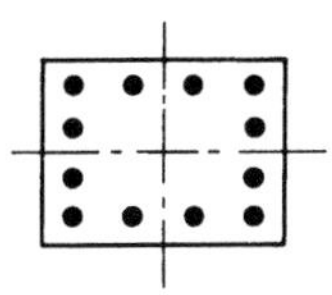

Figure 3.12. Cross-section of reinforced concrete column.

$$\mu_R = r_c + \sum_{i=1}^{12} \mu_{R_i} \tag{3.38}$$

and

$$\sigma_R = \left(\sum_{i=1}^{12} \sigma_{R_i}^2\right)^{\frac{1}{2}} \tag{3.39}$$

Assuming further that the various R_i are also identically distributed normal variables, N(100, 20) with units of kN, and that $r_c = 500$ kN, then

$$\mu_R = 500 + 12 \times 100 = 1700 \text{ kN} \quad \text{and} \quad \sigma_R = 69.28 \text{ kN}$$

Since R is also normally distributed in this case, the value of R which has a 99.99% chance of being exceeded is thus

$$\mu_R + \Phi^{-1}(0.0001)\sigma_R = 1700 - 3.719 \times 69.28 = 1442 \text{ kN}$$

This total load-carrying capacity corresponds to an *average* load-carrying capacity of (1442 − 500)/12 = 78.5 kN for the individual reinforcing bars, i.e. only 1.07 standard deviations below the mean.

For this type of structural configuration (in fact, a parallel ductile structural system in the reliability sense - see chapter 7) in which the structural strength is governed by the average strength of the components, it can be anticipated from the above - although it will not be formally proved here - that the reliability of the structure is not sensitive to the extreme lower tails of the strength distributions of the components. Hence the lack of availability of statistical data on extremely low strengths is not too important, for such cases.

Finally, it should be emphasised that these conclusions are based on the assumption that the various R_i are statistically independent.

Exercise 3.3. Given that the column discussed in example 3.3 is subjected to an axial load of 1500 kN, calculate the probability that this load exceeds the load-carrying capacity. Recalculate the probability under the assumption that the various R_i are mutually fully correlated ($\rho = +1$).

3.4.2 Choice of distributions for resistance variables

It has already been mentioned that unless experimental data are obtained from an effectively homogeneous source, formal attempts to fit standard forms of probability distribution to the data are hardly worthwhile. When data from two or more sources are present in a single sample, the overall shape of the cumulative frequency distribution is likely to depend as much, if not more, on the relative number of observations from each source than on the actual, but unknown, probability distribution of each sub-population. Extreme caution should therefore be exercised if the type of probability distribution is to be chosen on the basis of sample data alone.

A preferable approach is to make use of physical reasoning about the nature of each particular random variable to guide the choice of distribution. A number of limiting cases will now be studied.

The normal (Gaussian) distribution: As discussed in chapter 2, this is one of the most important probability distributions. It arises whenever the random variable of interest X is the *sum* of n identically distributed independent random variables, Y_i, irrespective of the probability distribution of Y_i, provided the mean and variance of Y_i are finite.

Formally, if $Y_1, Y_2, \ldots, Y_n$ are independent identically distributed random variables with finite mean μ_Y and finite variance σ_Y^2, and if $X = Y_1 + Y_2 + \ldots + Y_n$, then as $n \to \infty$

$$P(\alpha < \frac{X - n\mu_Y}{\sigma_Y \sqrt{n}} \leqslant \beta) \to \Phi(\beta) - \Phi(\alpha) \tag{3.40}$$

for all $\alpha, \beta\,(\alpha < \beta)$, and where Φ is the standard normal distribution function. This is known as the *central limit theorem.*

Provided a further set of conditions hold, the central limit theorem also applies to the sum of independent variables which are not identically distributed. The rate at which the sum tends to normality depends in practice on the presence of any dominant non-normal components.

It is therefore clear that any structural member whose strength is a linear function of a number of independent random variables may in general be considered to be governed by the normal law.

: **Example 3.6.** Consider again the reinforced concrete column discussed in example 3.5. Since
: the strength of the concrete is assumed known and the strengths of the reinforcing bars have
: been assumed to be independent, it may be concluded that the load-carrying capacity of the
: column R is normally distributed. (Whether this is true in practice clearly depends on a num-
: ber of other factors and whether these assumptions are valid).

It is sometimes argued that the normal distribution should not be used to model resistance variables because it gives a finite probability of negative strengths. However, this apparent criticism can be assumed to be relatively unimportant if the strength of a component can be considered to be the sum of a number of independent random variables, thereby invoking the central limit theorem.

The logarithmic normal distribution: The logarithmic normal (or log-normal) distribution is frequently used for modelling resistance variables and has the theoretical advantage of precluding negative values. The mathematical form and parameters of the log-normal distribution were discussed in chapter 2 (equation (2.51)). The log-normal distribution arises naturally as a limiting distribution when the random resistance X is the *product* of a number of independent identically distributed component variables, i.e.

$$X = Z_1 Z_2 \ldots Z_n = \prod_{i=1}^{n} Z_i \tag{3.41}$$

Clearly Y given by

$$Y = \ell nX = \ell nZ_1 + \ell nZ_2 + \ldots + \ell nZ_n = \sum_{i=1}^{n} \ell nZ_i \tag{3.42}$$

tends to normality as $n \to \infty$, following the central limit theorem, regardless of the probability distribution of ℓnZ_i. The probability distribution of X, therefore, tends towards the log-normal, as n increases.

Whether X may be regarded as a log-normal random variable in any practical situation in which X is the product of a number of random variables depends on the circumstances. The log-normal distribution is, however, used very widely in reliability studies.

Example 3.7. Many friction problems are governed by relationships of the form

$$P = ke^{\mu\alpha} \tag{3.43}$$

where k, μ and α are variables.

It is therefore to be expected that strength parameters which are affected by friction, (e.g. the shear strength of cohesionless soils, cables, etc.) will tend to be log-normally distributed, since spatial variations in the coefficient of friction μ will give rise to expressions of the form

$$P = ke^{\mu_1\alpha_1}\, e^{\mu_2\alpha_2} \ldots e^{\mu_j\alpha_j} \tag{3.44}$$

The Weibull distribution: This distribution is used quite frequently to model the distribution of the strength of a structural component whose strength is governed by size of its largest defect. If it is assumed that certain components, such as welded joints, contain a large number of small defects and that the severity of these defects is distributed in an appropriate manner, the distribution of the component strength approaches the Weibull distribution. As discussed in section 3.3.3 it is one of the so-called asymptotic extreme value distributions. Its density function is given in equations (2.55) and (2.56).

Other distributions: A number of other common distributions exist which may on occasions be useful for modelling the uncertainty in resistance variables - for example, the rectangular, beta, gamma and t-distributions. For information of these distributions the reader should consult a standard text, e.g. [3.5].

3.5 MODELLING OF LOAD VARIABLES - MODEL SELECTION

3.5.1 General remarks

The term *load* is generally understood to mean those *forces* acting on a structure which arise from external influences - principally the effects of gravity, and aerodynamic and hydrodynamic effects, e.g. structural self-weight, superimposed loads, snow, wind and wave loads. The term *action* is now often used as a more general description to include both loads and *imposed deformations.* Examples of the latter are dimensional changes arising from temperature effects and differential settlement. Both loads and imposed deformations give rise to sets of *action-effects* (often loosely referred to as load-effects) within a structure, e.g. bending moments and shear forces.

Unlike resistance variables, most of which change very little during the life of a structure, loads and other actions are typically time-varying quantities. The main exception of course is the self-weight of permanent structural and non-structural components. As mentioned earlier, time-varying quantities are best modelled as stochastic processes, but discussion of this topic is postponed to chapters 9 and 10.

It is often helpful to classify loads and other actions in accordance with the following three attributes [3.9]. Each load or action can be described as

- permanent or variable
- fixed or free
- static or dynamic

These three independent attributes relate to the nature of the action with respect to

- its variability in magnitude with time
- its variability in position with time
- the nature of the induced structural response

Thus the load imposed by vehicles on a lightly-damped long-span bridge could be described as being *variable*, *free* and *dynamic*. In general, loads and actions cannot be sensibly classified without a knowledge of the structure on which they are acting. For any particular action and structure, the attributes listed above also govern the nature of the structural analysis that must be undertaken.

To some degree nearly all loads could be considered to be variable, free and dynamic, but whether each is classified as such depends on the response of the structure to the loading.

: **Example 3.8**. Consider a steel bridge loaded solely by a sequence of partially-laden vehicles.
: As far as the imposed loads are concerned, the probability of failure of the bridge by a sim-
: ple plastic collapse mechanism depends only on the weight of the heaviest vehicle (assuming
: that only one vehicle can be on the bridge at any one time). However, the probability of
: failure by fatigue will also depend on (a) the weights of the other vehicles and (b) whether
: the individual vehicles induce any appreciable dynamic response. Clearly, there is only one
: source of loading, but the way in which it is classified and modelled is dictated by the fail-
: ure mode being analysed.

It should be noted that the preceding classification applies both to the actions themselves and to the mathematical models that are used to represent them.

A further distinction that should be made is between loading models used for the purposes of normal (deterministic) design and those required for structural reliability analysis. To take the simplest case, although a permanent fixed load is considered to be an action which does not vary with time or in position, it must generally be classed as an uncertain quantity for the purposes of reliability analysis, since in general its magnitude will not be known. It must therefore be modelled as a random variable. However, for deterministic design purposes it can be represented by a single specified constant.

It will not have escaped the attention of the reader that the modelling of loads and actions requires a certain degree of subjective judgement. The same is true for resistance variables. This should not, however, be seen as a limitation, since the aim is not to produce a perfect image of reality (an impossible task), but to develop a mathematical model of the phenomenon which embodies its salient features and which can be used to make optimal design decisions using the data available.

Finally, it should be noted that some »loads» act in a resisting capacity for some failure modes - for example, a proportion of the self-weight of the structure in most over-turning problems. In such cases, these »loads» are strictly resistance variables from a reliability viewpoint. They are generally easy to identify.

3.5.2 Choice of distributions for loads and other actions

We now consider the process of defining appropriate random variables and their associated probability distributions to model single loads and other actions. The modelling of combinations of loads is discussed in chapter 10. As in the case of resistance variables, the process consists of three distinct steps

- precise definition of the random variables used to represent the uncertainties in the loading
- selection of a suitable type of probability distribution for each random variable, and
- estimation of suitable distribution parameters from available data and any prior knowledge.

In many respects the first step is both the most important and the most difficult to decide upon in practice.

: **Example 3.9.** Consider the modelling of the asphalt surfacing on a long-span steel bridge.
: Should the surfacing be treated as a permanent or a variable load? How should spatial varia-
: tions in this load be taken into account? Should variations in density as well as variations
: in thickness be modelled? What is the probability that an additional layer of asphalt will be
: placed on the bridge without removal of the original surfacing and how should this be al-
: lowed for?

: These are typical of the questions that must be asked in any realistic load modelling prob-
: lem. They are also questions that can only be sensibly answered when the precise purpose
: of the proposed reliability analysis is known.

The second step of selecting a suitable probability distribution for each random variable can rarely be made on the basis of sample data alone and as in the case of resistance variables physical reasoning must be used to assist in this process. Some general guidelines are given below. The third step of evaluating suitable distribution parameters is discussed in section 3.6.

Permanent loads: The total permanent load that has to be supported by a structure is generally the sum of the self-weights of many individual structural elements and other parts. For this reason (see page 53) such loads are well represented by normal probability distributions. Whether the weights of individual structural elements can also be assumed to be normally distributed depends on the nature of the processes controlling their manufacture.

When the total permanent load acting on a structure is the sum of many independent components, it can easily be shown that the coefficient of variation of the total load is generally much less than those of its components.

: **Exercise 3.4.** Given that the total load on a foundation is the sum of n independent but
: identically distributed permanent loads P_i, show that the coefficient of variation of the
: total load is only $1/\sqrt{n}$ times that of the individual loads.

Variable loads: For continuous time-varying loads which can be uniquely described by a single quantity X (e.g. a magnitude), one can define a number of different but related probability distribution functions. The most basic is the so-called *arbitrary-point-in-time* or *first-order distribution* of X.

Let $x(t')$ be the magnitude of a single time-varying load $X(t)$ at time t'. For example, see figure 3.13 which shows a continuous state/continuous time stochastic process. Then F_X is the arbitrary-point-in-time distribution of $X(t)$ and is defined by

$$F_X(x) = P(X(t') \leqslant x) \tag{3.45}$$

where t' is any randomly selected time. The corresponding density function f_X is also illustrated in figure 3.13. F_X may take on a wide range of form and depends on the nature of $X(t)$ - i.e. whether $X(t)$ is a deterministic or stochastic function of time, whether the load can assume both negative and positive values, etc.

Example 3.10. If the load $X(t)$ has a deterministic time-history given by

$$x(t) = \hat{x}\sin(\omega t)$$

i.e. $x(t)$ is a sinusoidally-varying force of known amplitude $\hat{x}$, then it can be shown that

$$f_X(x) = \begin{cases} 0 & , \quad x < -x \\ \dfrac{1}{\pi\sqrt{\hat{x}^2 - x^2}} & , \quad -\hat{x} \leqslant x \leqslant \hat{x} \\ 0 & , \quad x > \hat{x} \end{cases}$$

which is a U-shaped distribution.

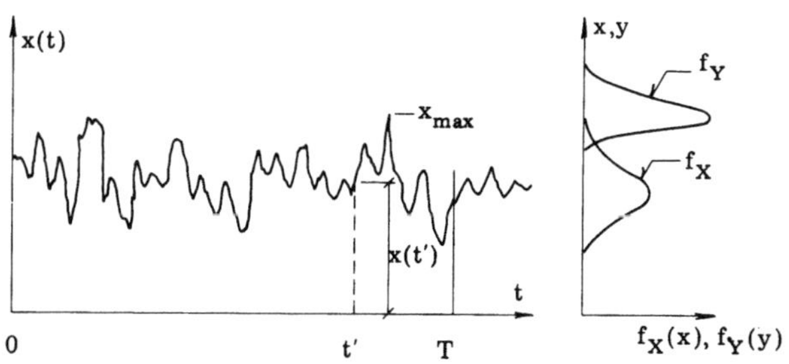

Figure 3.13. Illustration of continuous time-varying load.

: **Example 3.11.** Variations due to waves in the surface elevation of the sea X(t) at any
: fixed point remote from the shore can be shown to have a first-order distribution F_X
: which approximates very closely to the normal distribution (for periods of time in which
: the sea-state can be assumed to be stationary).

However, when dealing with single time-varying loads and so-called first passage problems (i.e. when failure occurs if and only if the load exceeds some threshold value), the form of the arbitrary-point-in-time distribution is not of immediate relevance. The random variable which is of importance is the magnitude of the largest extreme load that occurs during the reference period T for which the reliability is to be determined. The latter might be the specified design life or any other period of time.

If the loading process X(t) can be assumed to be ergodic (see chapter 9), the distribution of the largest extreme load can be thought of as being generated by sampling the values of x_{max} from successive reference periods T. If the values of x_{max} are represented by the random variable Y, then F_Y is the distribution function of the largest extreme load. The corresponding density function f_Y is illustrated in figure 3.13 and can be compared with the density function of the arbitrary-point-in-time distribution f_X.

Since, for a continuous loading process, the largest extreme load that occurs during any reasonably long reference period T corresponds to the largest of a finite number of peak loads, it can be seen from sections 3.2 and 3.3 that the probability distribution of the largest extreme is likely to be very closely approximated by one of the asymptotic extreme-value distributions. These distributions are frequently used for representing the maxima of time-varying loads. It should be noted, however, that the precise form and parameters of the extreme-value distribution depend very strongly on the autocorrelation function of the loading process X(t). The concept of autocorrelation is discussed in chapter 9.

For the present purposes it is sufficient to state that the maxima of time-varying loads can in most cases be represented by one of the asymptotic extreme-value distributions, with parameters estimated by one of the techniques given in section 3.6.

3.6 ESTIMATION OF DISTRIBUTION PARAMETERS

It is assumed that the selection of the types of probability distribution for the various load and resistance variables has been made using the approaches and methods of reasoning discussed previously. The problem now is to estimate suitable numerical values for the parameters of these distributions using available data. For single distributions this requires just one set of data, but for the more complex mixed distribution models such as shown in equation (3.34) various sets of data are clearly required.

The overall process of parameter estimation consists of

- initial inspection of the data
- application of a suitable estimation procedure
- final model verification.

It cannot be emphasised too strongly that the blind application of statistical procedures can lead to very misleading results and that an initial inspection of the available data should always be undertaken before any formal calculations are made.

Let us consider the practical problem of estimating the parameters of a single distribution function from a single sample of experimental data. The first step is to check the data for obvious inconsistencies and errors. Manually recorded or copied data have a high probability of containing at least some transcription errors. These should be eliminated if possible. The second step is to plot the data in the form of a histogram to check for *outliers* and to confirm that its shape does not deviate markedly from the shape of the density function being fitted. If the histogram is clearly bi-modal when a uni-modal distribution is being fitted to the data or if the sample appears to be truncated when the variable is assumed to be unbounded, checks on the data source are clearly required. Inconsistencies are often found to arise when the set of data has been obtained from experimental test programmes in more than one laboratory. Such lumping of data is often necessary when the sample sizes would otherwise be very small, but this should be avoided if possible. Checks on the consistency of the means and variances of the various subsamples (see for example [3.5]) should generally be undertaken when practicable.

The next step is to estimate the parameters of the selected distribution using one or more of the techniques described in section 3.6.1 below. The basic methods are

- the method of moments
- the method of maximum likelihood
- various graphical procedures
- use of order statistics.

The last step is to check that the sample data are well modelled by the chosen distribution and parameters. Methods for doing this are briefly reviewed in section 3.6.2.

3.6.1 Techniques for parameter estimation

This is a large subject in itself and only a brief description is possible here. Readers unfamiliar with the various concepts should also study a specialist text [3.11], [3.5], [3.8].

It is assumed in the following that the distribution function is known or has been postulated and that its parameters are now to be estimated. Depending on the distribution type, one, two, three or more parameters will be involved. The general procedure is to obtain estimates of these unknown parameters in terms of appropriate functions of the sample values. The word *estimate* is used in this context advisedly. It should be clear that because of the random nature of the variable no sample, however large, is completely representative of the source from which it derives; and indeed, small samples may be markedly unrepresentative. For example, in a random sample of 10 independent observarions drawn from a normal distribution, there is a probability of approximately 1:1000 that by chance *all* observations will be greater than the mean. Any attempt to estimate the parameters (μ, σ) of the parent distribution from this particular sample will result in considerable error. This difficulty cannot be escaped, but the probability of large errors occurring decreases as the sample size increases.

In essence there are two types of estimates for distribution parameters that can be obtained - *point estimates* and *interval estimates.* A point estimate is a single estimate of the parameter whereas interval estimates allow certain additional confidence or probability statements to be made. In this section only point estimates will be discussed.

The different techniques of parameter estimation summarised below correspond to the use of different functions of the sample data and give rise to different *estimators* for the parameters. A number of desirable properties which characterize »good» estimators are *unbiasedness, efficiency* and *consistency*. (For a precise definition of these terms, see for example, [3.11]). No estimator, however, has all these properties and in practice the choice of estimator is governed by the particular requirements of the problem, or expediency.

Method of moments: Let the variable of interest X have a probability density function f_X, with parameters $\theta_1, \theta_2, \ldots, \theta_k$. From equation (2.35) the jth moment of X is given by

$$\zeta_j = E[X^j] = \int_{-\infty}^{\infty} x^j f_X(x)dx \tag{3.46}$$

Since f_X is a function of the k parameters $\theta_1, \theta_2, \ldots, \theta_k$, the right hand side of equation (3.46) is also a function of the same k parameters and ζ_j may be expressed as

$$\zeta_j = \zeta_j(\theta_1, \theta_2, \ldots, \theta_k) \tag{3.47}$$

Using equation (3.46) to generate the first k moments ζ_j we obtain k equations in the k unknown distribution parameters θ_j.

If we now consider a random sample of the variable X of size n with values $(x_1, x_2, \ldots, x_n)$ the equivalent sample moments are given by

$$m_j = \frac{1}{n} \sum_{i=1}^{n} (x_i)^j \tag{3.48}$$

Finally, the moment estimators $\hat{\theta}_j$, j = 1, ... , k for the k unknown distribution parameters θ_j may be obtained by equating the moments of X, ζ, and the sample moments m, i.e.

$$\zeta_j = m_j \quad , \quad j = 1, \ldots, k \tag{3.49}$$

Example 3.12. Let X be a normally distributed random variable, having parameters μ and σ. The density function given by equation (2.45) is

$$f_X(x) = \frac{1}{\sigma\sqrt{2\pi}} \exp(-\frac{1}{2}(\frac{x-\mu}{\sigma})^2) \tag{3.50}$$

Assume that a random sample of n observations of X has been obtained, $(x_1, x_2, \ldots, x_n)$. The moment estimators for μ and σ^2 are now determined as follows.

Using equation (3.46) it can be shown that

$$\zeta_1 = E[X] = \mu \tag{3.51}$$

$$\zeta_2 = E[X^2] = \mu^2 + \sigma^2 \qquad (3.52)$$

The equivalent sample moments are

$$m_1 = \frac{1}{n} \sum_{i=1}^{n} x_i \qquad (3.53)$$

$$m_2 = \frac{1}{n} \sum_{i=1}^{n} x_i^2 \qquad (3.54)$$

Hence by equating terms, the estimators $\hat{\mu}$ and $\hat{\sigma}^2$ for the parameters μ and σ^2 may be obtained from

$$\hat{\mu} = \frac{1}{n} \sum_{i=1}^{n} x_i \qquad (3.55)$$

and

$$\hat{\sigma}^2 = \frac{1}{n} \sum_{i=1}^{n} x_i^2 - \hat{\mu}^2 \qquad (3.56)$$

giving

$$\hat{\sigma}^2 = \frac{1}{n} \left(\sum_{i=1}^{n} x_i^2 - \left(\sum_{i=1}^{n} x_i \right)^2 / n \right) \qquad (3.57)$$

Alternatively $\hat{\sigma}^2$ may be expressed as

$$\hat{\sigma}^2 = \frac{1}{n} \sum_{i=1}^{n} (x_i - \hat{\mu})^2 \qquad (3.58)$$

where $\hat{\mu}$ is the sample mean. However, the form given in equation (3.57) is in fact preferable from a computational point of view.

$\hat{\mu}$ and $\hat{\sigma}^2$ given by equations (3.56) and (3.57) are thus the moment estimators of μ and σ^2, respectively. It should be noted, however, that the best *unbiased* estimator of σ^2 is not $\hat{\sigma}^2$ but $S^2 = (n/(n-1))\,\hat{\sigma}^2$.

Method of maximum likelihood: This method is generally more difficult to apply than the method of moments, often involving iterative calculations, but maximum likelihood estimators of distribution parameters can be shown to have a number of desirable properties [3.11].

Let the variable of interest X have a probability density function f_X with unknown parameters $\bar{\theta} = (\theta_1, \theta_2, \ldots, \theta_k)$ that are to be determined. Assume, in addition, that a particular random sample $(x_1, x_2, \ldots, x_n)$ of the random variable X has been obtained. The *likelihood function* of this sample is defined as

$$L(\bar{\theta} \mid x_1, x_2, \ldots, x_n) = \prod_{i=1}^{n} f_X(x_i \mid \bar{\theta}) \qquad (3.59)$$

L expresses the relative likelihood of having observed the sample as a function of the parameters $\bar{\theta}$. Referring to equation (2.68) it can be seen that the right hand side of equation (3.59) is the joint density function $f_{X_1, X_2, \ldots, X_n}(x_1, x_2, \ldots, x_n \mid \bar{\theta})$ of a sample with n elements $x_1, x_2, \ldots, x_n$ taken at random from the variable X. In this case, however, it is the sample values $x_1, x_2, \ldots, x_n$ that are known and the parameters $\bar{\theta}$ that are treated as variables.

The *maximum-likelihood estimators* $\hat{\bar{\theta}}$ of the parameters $\bar{\theta}$ are defined as the values of $\bar{\theta}$ that maximize L, or, equivalently and more conveniently, the logarithm of L. The evaluation of $\hat{\bar{\theta}}$ thus requires the solution of the set of k equations

$$\sum_{i=1}^{n} \frac{\partial}{\partial \theta_j} \log(f_X(x_i | \bar{\theta})) = 0 \quad , \quad j = 1, 2, \ldots, k \tag{3.60}$$

taking due account of any constraints (e.g. $0 < \hat{\sigma} < \infty$ for the parameter σ of a normal distribution).

: **Exercise 3.5.** Derive the maximum-likelihood estimators $\hat{\mu}$ and $\hat{\sigma}$ for the parameters μ
: and σ of a normal distribution. Show that for this distribution, these estimators are the
: same as those obtained by the method of moments.

Graphical procedures: For most simple probability distributions, it is possible to plot the cumulative distribution function F_X for different values of the variable x as a straight line, simply by pre-selecting an appropriate plotting scale or type of probability paper. See, for example, figures 3.4 and 3.10.

: **Example 3.13.** Let the random variable X have a 2-parameter Weibull distribution with
: parameters β and k and distribution function
:
: $$F_X(x) = 1 - \exp(-(x/k)^{\beta}) \quad , \quad x \geqslant 0, \beta > 1, k > 0 \tag{3.61}$$
:
: Then, $z = \ell n(-\ell n(1 - F_X(x)))$ is a linear function of $y = \ell n x$, since
:
: $$\ell n(-\ell n(1 - F_X(x))) = \beta \ell n x - \beta \ell n k \tag{3.62}$$
:
: The variables x and y therefore plot as a straight line on natural scales. Equivalent scales
: in the original quantities $F_X(x)$ and x can therefore be constructed.

If we now obtain a random sample of size n from a known type of distribution function F_X, but with unknown parameters $\bar{\theta}$, the cumulative frequency distribution for the sample can be expected to plot as a straight line if the appropriate plotting scale is used. It is usual to order the elements of the sample $(x_1, x_2, \ldots, x_n)$ to obtain the sequence $x_1^n, x_2^n, \ldots, x_i^n, \ldots, x_n^n$, where x_1^n is the smallest value and x_i^n is the ith largest value called the ith *order statistic*. It will be recalled that the probability distribution function for the random variable X_i^n was derived in section 3.2.

One estimate of the cumulative distribution function $F_X(x_i)$ (i.e. the particular value of F_X for $X = x_i$) is thus i/n, but preferable estimates are $i/(n+1)$ or $(i - 1/2)/n$, since for most distribution types they can be shown to be less biased. The cumulative frequency diagram is therefore obtained by plotting the points $(x_i, i/(n+1))$ using scales appropriate to the type of distribution function. It should be noted, however, that some random deviations from a straight line are to be expected, particularly for points at each end of the line.

For one and two-parameter probability distributions, estimates of the distribution parameters can then be obtained by drawing the »best» straight line through the plotted points either by eye or using a formal least-squares method. In both cases, it is the sum of the horizontal squared deviations from the line which should be minimised, not the vertical (assuming the axes are chosen as shown in figure 3.4). Finally, the estimates of the distribution parameters are obtained from the slope and position of the best straight line.

Use of order statistics: The graphical method discussed above is in fact a simple application of order statistics. A detailed discussion of this subject is beyond the scope of this book. The general approach in estimating the parameters of distributions of known type is to use sets of coefficients or weighting factors in conjunction with the order statistics to obtain estimates of the parameters. The coefficients are chosen to give unbiased and highly efficient estimates for samples of particular size. The approach was first used and has subsequently been further developed by Lieblein [3.10] for extreme-value distributions. See also [3.11]. This approach should not be neglected in any serious application of these methods.

3.6.2 Model verification

The final stage in the process of distribution selection and parameter estimation should be model verification. For situations in which only one set of data and no other information is avaiable, the approach is straightforward. The simplest method is to check whether the sample data plot as a reasonable straight line on the appropriate probability paper. If the distribution parameters have been estimated graphically, this step will have been taken as part of the estimation procedure. The sample data shown in figure 3.10 may be considered to be a »good» straight-line plot. Alternatively, a formal *goodness-of-fit test*, such as the χ^2 test or the Kolmogorov-Smirnov test may be employed to ascertain the level of probability at which it is possible to reject the null hypothesis that »the random variable X has a particular distribution function with certain stated parameters». Such tests are widely described, e.g. [3.5], and will not be given here.

In many structural reliability problems, however, the basic variables are best described by mixed distribution models for which the tests described above are not applicable. In other cases, the analyst may prefer to use some prescribed distribution type to model a basic variable, e.g. a log-normal distribution to model a resistance variable, even though over the limited range of available data some other distribution type may in fact give a better fit. The formal use of goodness-of-fit tests in the context of structural reliability theory is therefore often limited.

3.7 INCLUSION OF STATISTICAL UNCERTAINTY

As mentioned previously, the analyst is often faced with the problem of having insufficient data for one or all of the basic variables which affect the structural reliability. Let us assume, however, that there are good a priori reasons for assuming that a particular basic random variable X is governed by a particular type of probability distribution. The problem arises therefore of selecting the values of the parameters $\overline{\theta}$ for that distribution.

One approach is to use single point estimates for the parameters - for example, the maximum-likelihood estimates - and to ignore the additional statistical uncertainty that arises when there are too few data. This approach may not be too unconservative since any non-homogeneity in the data will tend to artificially enhance the variance. A better approach is to include the statistical uncertainty in the parameters within the distribution of X itself, in terms of what is known as the predictive distribution of X.

If the probability density function of the random variable X, for known parameters $\overline{\theta}$ is written as $f_X(x|\overline{\theta})$ then the predictive density h_X for uncertain $\overline{\theta}$ is given by

$$h_X(x) = \int_{\overline{\theta}} f_X(x|\overline{\theta}) f''_{\overline{\theta}}(\overline{\theta}|\overline{z}) d\overline{\theta} \tag{3.63}$$

where $f''_{\overline{\theta}}(\overline{\theta}|\overline{z})$ is the posterior probability density for $\overline{\theta}$ given a set of data $\overline{z} = (z_1, z_2, \ldots, z_z)$. $f''_{\overline{\theta}}(\overline{\theta}|\overline{z})$ can be obtained from Bayes theorem (see equation (2.24)) which can be expressed as

$$f''_{\overline{\theta}}(\overline{\theta}|\overline{z}) = N L(\overline{\theta}|z_1, z_2, \ldots, z_n) f'_{\overline{\theta}}(\overline{\theta}) \tag{3.64}$$

where

$L(\overline{\theta}|\overline{z})$	is the likelihood of $\overline{\theta}$ given the observation $\overline{z}$, and
$f'_{\overline{\theta}}(\overline{\theta})$	is the prior density of $\overline{\theta}$, before obtaining the data, and
N	is a normalising constant.

For further information the reader is referred to Aitchison and Dunsmore [3.1].

BIBLIOGRAPHY

[3.1] Aitchison, J. & Dunsmore, I. R.: *Statistical Prediction Analysis.* Cambridge University Press, 1975.

[3.2] Baker, M. J.: *Variability in the Strength of Structural Steels - A Study in Structural Safety. Part I: Material Variability.* CIRIA Technical Note 44, September 1972.

[3.3] Baker, M. J. & Wickham, A. H. S.: *An Examination of the Within-Cast Variability of the Yield Strength of Hot-Rolled High-Yield Reinforcing Bars.* Imperial College. Department of Civil Engineering Report SRRG/3/80, September 1980.

[3.4] Bannister, J. L.: *Steel Reinforcement and Tendons for Structural Concrete. Part I: Steel for Reinforced Concrete.* Concrete, Vol. 2, July 1968.

[3.5] Benjamin, J. R. & Cornell, C. A.: *Probability, Statistics and Decision for Civil Engineers.* McGraw-Hill, 1970.

[3.6] Edlund, B. & Leopoldson, U.: *Scatter in Strength of Data of Structural Steel.* Publication S72:4. Dept. Struct. Eng., Chalmers University of Technology.

[3.7] Ferry Borges, J. & Castanheta, M.: *Structural Safety.* Laboratorio Nacional de Engenharia Civil, Lisbon, 1971.

[3.8] Gumbel, E. J.: *Statistics of Extremes.* Columbia University Press, 1958.

[3.9] Joint Committee on Structural Safety, CEB - CECM - CIB - FIP - IABSE - IASS - RILEM: *General Principles on Reliability for Structural Design.* International Association for Bridge and Structural Engineering. Reports of the Working Commissions. Volume 35, 1981.

[3.10] Lieblein, J.: *Efficient Methods of Extreme-Value Methodology.* U.S. Department of Commerce, National Bureau of Standards. Report NBSIR 74-602, 1974.

[3.11] Mann, N. R., Schafer, R. E. & Singpurwalla, N. D.: *Methods for Statistical Analysis of Reliability and Life Data.* John Wiley & Sons, 1974.

[3.12] Mathieu, H.: *Manuel »Sécurité des Structures».* Comité Euro-International du Béton. Bulletin d'Information Nos 127 and 128, 1980.

[3.13] Mayne, J. R.: *The Estimation of Extreme Winds.* Journal of Industrial Aerodynamics, Vol. 4, 1979.

[3.14] Mitchell, G. R. & Woodgate, R. W.: *A Survey of Floor Loads in Office Buildings.* Construction Industry Research and Information Association, Report 25, London, 1970.

[3.15] Sentler, L.: *A Stochastic Model for Live Loads on Floors in Buildings.* Div. of Building Technology, Lund Institute of Technology, Report 60, 1975, Lund, Sweden.

Chapter 4

FUNDAMENTALS OF STRUCTURAL RELIABILITY THEORY

4.1 INTRODUCTION

Structural reliability analysis differs in many important ways from reliability analysis as practised in, for example, the electronics and aero-space industries, in spite of the fact that the underlying probabilistic nature of the problems are the same. A clear understanding of these differences is helpful and they will be discussed later. Both branches have developed from classical reliability theory - a subject which evolved as a result of the increasing need for reliable electronic systems during the 1940's, initially for military applications.

The principles of reliability analysis have been applied to a very large class of problems, ranging from the design of control systems for complex nuclear and chemical plant to the design of specific mechanical and structural components, as well as more generally in the field of electronics and aero-space. Reliability analysis should not, however, be thought of as an isolated discipline as it is closely related to the theory of statistics and probability and to such fields as operations research, systems engineering, quality control engineering and statistical acceptance testing.

4.2 ELEMENTS OF CLASSICAL RELIABILITY THEORY

Classical reliability theory was developed to predict such quantities as the expected life, or the expected failure rate, or the expected time between breakdowns of mechanical and electronic systems, given some test or failure data for the system and/or its components. These predictions may then be used to answer such questions as: What is the probability that the actual life of a particular system will exceed the required or specified design life? Taking into account the consequences of system malfunctions, is it economic to increase the expected life by providing more component redundancy? What is the optimal period of time that should elapse before a system or component is inspected or replaced? There are many questions of this type and the nature of the system, the use to which it will be put and the consequences of failure dictate the type of analysis that should be undertaken.

Reliability function: Typically, however, the probability of failure of a system or component is a function of operating or exposure time; so that the reliability may be expressed in terms of the distribution function F_T of the variable T, the random time to failure. The *reliability function* $\mathcal{R}_T$ which is the probability that the system will still be operational at time t is given by

$$\mathcal{R}_T(t) = 1 - F_T(t) = 1 - P(T \leqslant t) = P(T > t) \tag{4.1}$$

If the density function f_T of the time to failure is known then $\mathcal{R}_T$ may be expressed as

$$\mathcal{R}_T(t) = 1 - \int_0^t f_T(\tau)d\tau = \int_t^\infty f_T(\tau)d\tau \tag{4.2}$$

In some circumstances there are good a priori reasons for selecting a particular form for f_T – for example, a Weibull distribution, whose distribution function takes the form

$$F_T(t) = 1 - \exp[-(\frac{t}{k})^\beta] \tag{4.3}$$

Substituting in equation (4.1) gives

$$\mathcal{R}_T(t) = e^{-(t/k)^\beta} \quad , \quad t \geqslant 0 \tag{4.4}$$

Expected life: A further property of a system or component which is of interest is the *expected life* or the expected time for which the system or component can be expected to operate satisfactorily. Referring to chapter 2, this is given by

$$E[T] = \int_0^\infty \tau f_T(\tau)d\tau = \int_0^\infty \mathcal{R}_T(t)dt \tag{4.5}$$

since, integrating by parts,

$$\int_0^\infty \mathcal{R}_T(t)dt = [t\mathcal{R}_T(t)]_0^\infty + \int_0^\infty t f_T(t)dt = E[T] \tag{4.6}$$

provided $\lim_{t \to \infty} [t\mathcal{R}_T(t)] = 0$.

Failure rates and hazard functions: Also of interest is a knowledge of how the rate of failure changes with time for any particular form of reliability function. The probability of failure within any given time interval $[t, t + \delta t]$ is the probability that the actual life T lies in the range t to $t + \delta t$ and is given by

$$P(t < T \leqslant t + \delta t) = F_T(t + \delta t) - F_T(t) = \mathcal{R}_T(t) - \mathcal{R}_T(t + \delta t) \tag{4.7}$$

The average rate at which failure occurs in any time interval $[t, t+\delta t]$ is defined as the *failure rate* and is the probability per unit time that failure occurs within the interval, given that it has not already occurred prior to time t, namely

$$\frac{\mathcal{R}_T(t) - \mathcal{R}_T(t+\delta t)}{\delta t\, \mathcal{R}_T(t)} \tag{4.8}$$

The *hazard function* is defined as the instantaneous failure rate as the interval δt approaches zero. It is normally denoted h and is given by

$$h(t) = \lim_{\delta t \to 0} \frac{\mathcal{R}_T(t) - \mathcal{R}_T(t+\delta t)}{\delta t\, \mathcal{R}_T(t)} = \frac{1}{\mathcal{R}_T(t)} \left[-\frac{d}{dt} \mathcal{R}_T(t)\right] = \frac{f_T(t)}{\mathcal{R}_T(t)} \tag{4.9}$$

The probability that a system or component, *which has already survived for a period of time t*, will fail in the next small interval of time dt is thus simply $h(t)dt$. Note that this probability is *not* $f_T(t)dt$.

The use of the hazard function is in indicating whether a system or component becomes progressively more or less likely to fail per unit time as time progresses. If it becomes progressively more likely to fail then clearly action should be taken to replace the component or system at some stage or to minimise the consequences of failure.

Example 4.1. Assume that some delicate instrument is destroyed by horizontal accelerations greater than 0.05g and that the time interval between major earthquakes can be assumed to be exponentially distributed. What is the form of the hazard function?

The exponential density function is given by

$$f_T(t) = \lambda e^{-\lambda t} \quad , \quad t \geqslant 0, \lambda > 0 \tag{4.10}$$

and where λ is a constant.
Hence

$$\mathcal{R}_T(t) = \int_t^{\infty} f_T(\tau) d\tau = \int_t^{\infty} \lambda e^{-\lambda\tau} d\tau = e^{-\lambda t} \tag{4.11}$$

and thus

$$h(t) = \frac{f_T(t)}{\mathcal{R}_T(t)} = \lambda \tag{4.12}$$

For this particular form of distribution of time to failure, the hazard function is a constant, and thus the probability of failure per unit time is independent of preceding events.

Choice of reliability function: In many areas of electronic, electrical and mechanical engineering, data are available for the time to failure of items such as relays, bearings, driveshafts and many other components. In fact data banks for this type of information exist in many industries. The »time to failure« may, however, be recorded in terms of number of cycles of operation, number of miles driven, etc., rather than in units of actual time.

With some types of component, large quantities of data may be available on time to failure and a distribution may be selected which gives the best fit to the data. Care must be taken, however, to ensure that the data are effectively homogeneous and do not include, for example, systematic changes with time.

Frequently, however, the data may be limited in number, in which case it is necessary to select the type of distribution a priori - from an understanding of the physical nature of the failure mechanism and/or from previous experience.

Systems: In general, the aim of design is to provide an assembly of components which when acting together will perform satisfactorily for some nominal design life either with or without maintenance. Frequently, time to failure data are available for the individual components, but are not available for the complete system. The reason for this is that complete systems are more difficult and costly to test and secondly because it is often the aim of the design to provide much higher reliability for the complete system than for the individual components, by providing a high level of component and sub-system redundancy.

In these cases it is necessary to compute the reliability function and hazard function for the system from a knowledge of the reliability functions of the components and a knowledge of how the components are inter-connected. In general, account should also be taken of the possibility of detecting and replacing components which have failed. It is, however, of the utmost importance to predict the existence of any adverse interactions between the components which may exist in the system but which have not affected the failure rates of the individual components when tested in isolation or under different conditions.

These problems will not be discussed in detail since they are different from the systems problems that are encountered in structural engineering. The latter are discussed in chapters 7 and 8.

4.3 STRUCTURAL RELIABILITY ANALYSIS

4.3.1 General

Attention will now be focused on the particular features of structural reliability analysis, but first, some of the principal differences between electronic/mechanical systems and structural systems will be examined. Most electrical, electronic and mechanical components and systems deteriorate during use as a result of elevated operating temperatures, chemical changes, mechanical wear, fatigue, overloading, and for a number of other reasons. Failure of a particular component may eventually occur for one of these reasons, or it may be caused indirectly as a result of the deterioration of some other part of the system. It is natural, therefore, to think in terms of the life of electrical and mechanical systems, although the precise time at which »failure» occurs may be difficult to establish since the definition of failure may be fuzzy or somewhat arbitrary.

In contrast to electronic/mechanical systems, structural systems tend not to deteriorate, except by the mechanisms of corrosion and fatigue, and in some cases may even get stronger - for example: the increase in the strength of concrete with time, and the increase in the strength of soils as a result of consolidation. Whilst basic data are available for the time to failure of electronic and mechanical components, no such information is available for structural components, because in general they do not fail in service. The problem is of a different nature.

Structures or structural components »fail» when they encounter an extreme load, or when a combination of loads causes an extreme load effect of sufficient magnitude for the structure to attain a »failure state»; this may be an ultimate or a serviceability condition. Part of the problem, therefore, is to predict the magnitude of these extreme events. The other part is to predict the strength or load-deflection characteristics of each structural component from the information available at the design stage. In other words, it is necessary to synthesise probabilistic models for the two parts of the problem, including, on the one hand, all the uncertainties affecting the loading, and, on the other, all the uncertainties affecting the strength or resistance of the structures. As discussed in chapter 3, statistical data may be available to quantify some of the sources of uncertainty and this information can be included in a suitable form. Other sources of uncertainty may have to be assessed subjectively.

A further difference between electronic/mechanical systems and structural systems is that whereas most structural systems are »one-off», all but the most complex electronic and mechanical systems are produced in considerable numbers and can be assumed to be nominally identical. The existence of this underlying population of nominally identical systems or components means that it is possible to interpret failure probabilities in terms of relative frequencies. This is less justified for civil engineering systems, and the probabilities that are determined should not be thought of in this way. They should be thought of as intermediate results in a decision-making process and with little or no absolute meaning of their own. As discussed in chapter 1, the calculated reliability or failure probability for a *particular structure* is *not* a unique property of that structure but a function of the reliability analyst's lack of knowledge of the properties of the structure and the uncertain nature of the loading to which it will be subjected in the future.

4.3.2 The Fundamental Case

For a simple structural member selected at random from a population with a known distribution function F_R of ultimate strength R in some specified mode of failure, the probability of failure P_f under the action of a single known load effect s is

$$P_f = P(R - s \leqslant 0) = F_R(s) = P(R/s \leqslant 1) \tag{4.13}$$

If the load effect S is also a random variable, with distribution function F_S, equation (4.13) is replaced by

$$P_f = P(R - S \leqslant 0) = \int_{-\infty}^{+\infty} F_R(x) f_S(x) dx \tag{4.14}$$

under the condition that R and S are statistically independent. Equation (4.14) is best understood by plotting the density functions of R and S, as shown in figure 4.1. It should be noted that R and S must necessarily have the same dimensions (e.g. loads and load-carrying capacities, or bending moments and flexural strength).

Equation (4.14) gives the total probability of failure P_f as the product of the probabilities of two independent events, summed over all possible occurrences; namely the probability P_1 that S lies in the range x,x + dx and the probability P_2 that R is less than or equal to x. It is clear that

$$P_1 = f_S(x)dx \tag{4.15}$$

and

$$P_2 = F_R(x) \tag{4.16}$$

Under these conditions the *reliability* $\mathcal{R}$ is the probability that the structure will survive when the load is applied, and is given by

$$\mathcal{R} = 1 - P_f = 1 - \int_{-\infty}^{+\infty} F_R(x) f_S(x) dx \tag{4.17}$$

From considerations of symmetry, it can be seen that the reliability may also be expressed as

$$\mathcal{R} = 1 - P_f = 1 - \int_{-\infty}^{+\infty} (1 - F_S(x)) f_R(x) dx \tag{4.18}$$

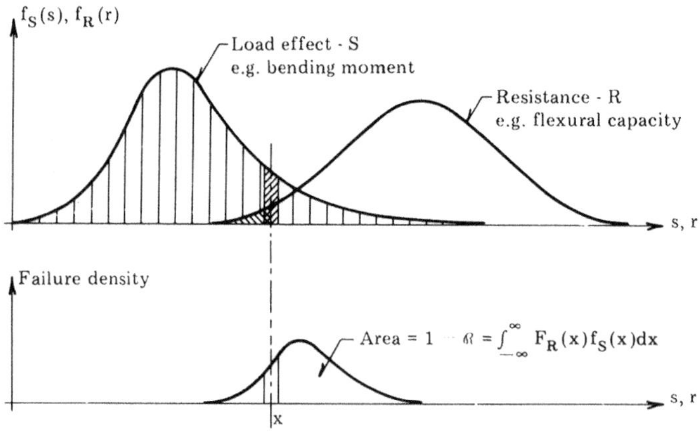

Figure 4.1

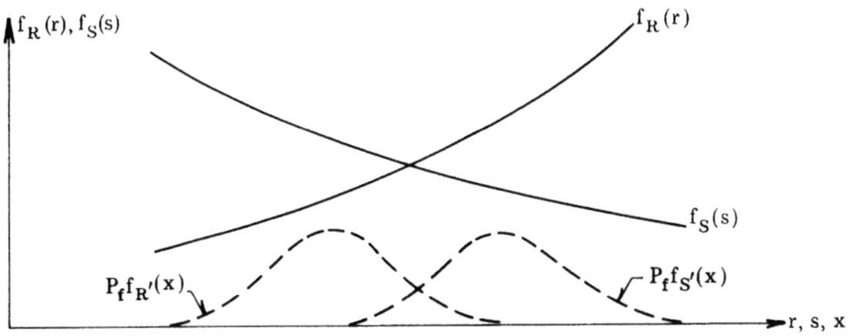

Figure 4.2.

Since in general it is not meaningful to have negative strengths, the lower limits of integration in equations (4.17) and (4.18) may in practice be replaced by zero.

It is also of interest to note that although equations (4.17) and (4.18) give identical numerical values for P_f they are in fact fundamentally different. Setting

$$f_{R'}(x) = F_R(x) f_S(x)/P_f \tag{4.19}$$

and

$$f_{S'}(x) = (1 - F_S(x)) f_R(x)/P_f \tag{4.20}$$

we obtain, respectively, the probability density function of the resistance R′ of structures in which failure occurs and the probability density function of the load effects S′ which have caused failure. These are not the same, since $r \neq s$ at failure, see figure 4.2.

It should be noted that P_f is *not* given by the area of overlap of the two density functions $f_R(r)$ and $f_S(s)$ in figure 4.1. This is a common misconception.

For the general case, closed-form solutions do not exist for the integrals in equations (4.17) and (4.18). There are, however, a number of special cases.

Example 4.2. If R and S are independent normally distributed variables, P_f may be expressed as

$$P_f = P(M \leqslant 0) \tag{4.21}$$

where

$$M = R - S \tag{4.22}$$

Thus

$$\mu_M = E[M] = \mu_R - \mu_S \tag{4.23}$$

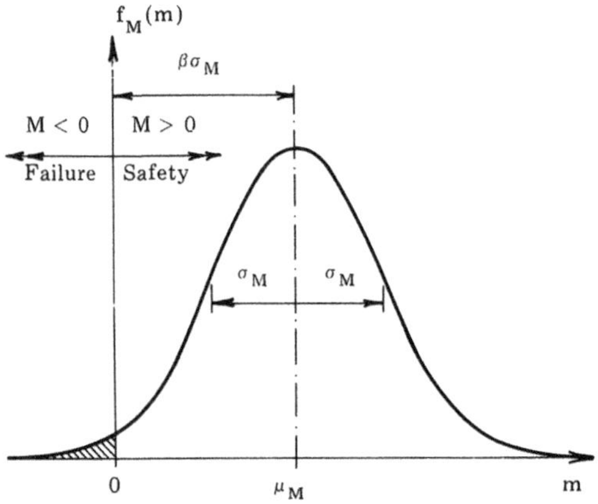

Figure 4.3. Illustration of the reliability index β.

and

$$\sigma_M^2 = \mathrm{Var}[M] = \sigma_R^2 + \sigma_S^2 \tag{4.24}$$

giving

$$\sigma_M = (\sigma_R^2 + \sigma_S^2)^{\frac{1}{2}} \tag{4.25}$$

Since R and S are normal, M, a linear function of R and S, is also normally distributed and $(M - \mu_M)/\sigma_M$ is unit standard normal, giving

$$P_f = \Phi\left(\frac{0 - \mu_M}{\sigma_M}\right) = \Phi\left(\frac{\mu_S - \mu_R}{(\sigma_S^2 + \sigma_R^2)^{\frac{1}{2}}}\right) \tag{4.26}$$

where Φ is the standard normal distribution function and μ_S, μ_R, σ_S and σ_R are respectively the means and standard deviations of random variables S and R.

The *reliability index* β may now be defined as the ratio μ_M/σ_M or the number of standard deviations by which μ_M exceeds zero. See figure 4.3.
Hence

$$P_f = \Phi\left(-\frac{\mu_M}{\sigma_M}\right) = \Phi(-\beta) \tag{4.27}$$

For the more general case where R and S are jointly normally distributed with correlation coefficient ρ, equation (4.27) still holds but σ_M is given by

$$\sigma_M = (\sigma_R^2 + \sigma_S^2 + 2\rho\sigma_R\sigma_S)^{\frac{1}{2}} \tag{4.28}$$

Example 4.3. If R and S are both log-normally distributed, P_f may be expressed as

$$P_f = P(M' \leqslant 1) \tag{4.29}$$

where

$$M' = R/S \tag{4.30}$$

Taking logarithms to the base e and putting $M = \ell n M'$ gives

$$M = \ell n R - \ell n S = A - B \tag{4.31}$$

where $A = \ell n R$ and $B = \ell n S$. Failure occurs when $M' < 1$, or when $M < 0$. But if R and S are lognormally distributed, A and B are normally distributed, so that M is also normally distributed. Thus

$$P_f = \Phi\left(\frac{\mu_B - \mu_A}{(\sigma_B^2 + \sigma_A^2)^{\frac{1}{2}}}\right) = \Phi\left(\frac{\mu_{\ell nS} - \mu_{\ell nR}}{(\sigma_{\ell nS}^2 + \sigma_{\ell nR}^2)^{\frac{1}{2}}}\right) \tag{4.32}$$

The properties of the lognormal distribution are such that if Y is lognormally distributed and $X = \ell n Y$, then

$$\sigma_X^2 = \ell n[V_Y^2 + 1] \tag{4.33}$$

$$\widetilde{m}_Y = \mu_Y \exp(-\frac{1}{2}\sigma_X^2) \tag{4.34}$$

and

$$\mu_X = \ell n(\widetilde{m}_Y) \tag{4.35}$$

where $\widetilde{m}_Y$ is the median of Y and V_Y is the coefficient of variation of $Y = \sigma_Y/\mu_Y$. Substituting in equation (4.32) gives

$$P_f = \Phi\left(\frac{\ell n\left(\frac{\mu_S}{\mu_R}\sqrt{\frac{V_R^2 + 1}{V_S^2 + 1}}\right)}{\sqrt{\ell n((V_R^2 + 1)(V_S^2 + 1))}}\right) \tag{4.36}$$

Convenient analytical expressions for P_f do exist for some other combinations of distributions of R and S. However, this simple representation of the problem is rarely of much practical use in structural reliability analysis, so we shall not consider them here. In general, equation (4.17) or (4.18) needs to be evaluated by numerical methods.

4.3.3 Problems Reducing to the Fundamental Case

In some simple situations, although R and S may each be functions of a number of other random variables, it may be possible by means of appropriate transformations to reduce the problem to the simple form. This is best illustrated by means of an example.

Example 4.4. A slender cylindrical column of diameter D is to be designed to carry a time-varying axial load P, the maximum value of which in any 50 year period may be assumed to be lognormally distributed with a mean of 250 kN and a coefficient of variation V_P = 0.25. The load-carrying capacity of the column may be assumed to be governed by the relationship

$$P_c = \frac{C\pi^3 ED^4}{64L^2}$$

but owing to the nature of the end restraints, there is considerable uncertainty in its effective length L. This uncertainty may be represented by modelling L as a lognormal variable with a mean of 4 m and a coefficient of variation $V_L = 0.15$. The other quantities C, the model uncertainty, and E, Young's modulus, may also be assumed to be lognormally distributed with the following parameters

$$\mu_C = 0.9 \qquad V_C = 0.1$$

$$\mu_E = 205 \text{ kN/mm}^2 \qquad V_E = 0.025$$

Assuming that there is no uncertainty associated with the diameter D, find the required value of D such that the column has a reliability of 0.9999 for a 50 year reference period.

This problem may be solved as follows. If the applied load is P, failure occurs when

$$P = P_c = \frac{C\pi^3 ED^4}{64\, L^2}$$

or re-arranging, when

$$D = \left(\frac{64\, L^2 P}{C\pi^3 E}\right)^{\frac{1}{4}}$$

Taking logarithms to the base e gives

$$\ell n(D) = \frac{1}{4}(\ell nP + \ell n 64 + 2\ell nL - \ell nC - 3\ell n\pi - \ell nE)$$

Replacing D, the actual diameter, by Θ, the variable diameter at which failure will just occur, and using the rules of expectation and variance for linear functions gives

$$E[\ell n\Theta] = \frac{1}{4}(E[\ell nP] + E[\ell n 64] + 2E[\ell nL] - E[\ell nC] - 3E[\ell n\pi] - E[\ell nE])$$

and

$$\mathrm{Var}[\ell n\Theta] = \frac{1}{16}(\mathrm{Var}[\ell nP] + 4\,\mathrm{Var}[\ell nL] + \mathrm{Var}[\ell nC] + \mathrm{Var}[\ell nE])$$

But from equation (4.33)

$$\mathrm{Var}[\ell nP] = \ell n[V_P^2 + 1] = \ell n[0.25^2 + 1] = 0.0606$$

Similarly, Var $[\ell nC] = 0.00995$, $\mathrm{Var}[\ell nE] = 0.00062$ and $\mathrm{Var}[\ell nL] = 0.02225$. Substituting these values yields

$$\mathrm{Var}[\ell n\Theta] = 0.010013$$

and thus $\sigma_{\ell n\Theta} = 0.10006$.

From equations (4.34) and (4.35)

$$E[\ell nP] = \ell n\,\mu_P - \frac{1}{2}\sigma_P^2 = \ell n(250) - 0.5 \times (0.0606) = 5.49115$$

: Similarly, $E[\ell n C] = -0.1103$, $E[\ell n E] = 5.3227$ and $E[\ell n L] = 8.2829$.
: Substitution yields, $E[\ell n \Theta] = 4.3923$. But $E[\ell n \Theta] = \mu_{\ell n \Theta} = \ell n(\widetilde{m}_{\Theta})$ from equation (4.35).
: Hence the median value of Θ, $\widetilde{m}_{\Theta} = \exp(4.3928) = 80.83$ mm. Thus the value of the dia-
: meter which will result in failure 50% of the time is 80.83 mm.

: However, the required reliability is 0.9999. The value of the diameter $D^{\star}$ which will pro-
: vide this reliability is found as follows:
:
: $\beta = \Phi^{-1}(0.9999) = +3.72$
:
: (from tables of the standard normal distribution function), and since $\ell n \Theta$ is normally dis-
: tributed
:
: $\ell n\, D^{\star} = \mu_{\ell n \Theta} + 3.72\sigma_{\ell n \Theta} = 4.7646$
:
: giving $D^{\star} = D = \exp(4.7646) = 117.3$ mm.
: The required diameter for a reliability of 0.9999 in a 50 year period is thus 117.3 mm.

4.3.4 Treatment of a Single Time-Varying Load

The situation discussed in section 4.3.2 was that of an uncertain load effect S applied once to a structure of uncertain resistance R. The example given above shows how R and S can be modelled by a number of component variables. It also shows how a single time-varying load may be treated. The correct approach is to use the probability distribution for the load W corresponding to the maximum intensity in the reference period T for which the reliability is being determined. In the preceding example, this was the 50 year extreme maximum distribution. These distributions were discussed in chapter 3. Provided the correct distribution is used, the time element may then be neglected in the reliability calculations.

When a reliability problem involves more than one time-varying load the problem is more complex (see chapter 10).

4.3.5 The General Case

Only in the simplest of structures can the reliability be expressed in terms of the two random variables R and S. This is mainly because R and S are not known or are not convenient mathematical expressions, e.g.

$$\begin{aligned} R &= \text{function(material properties, dimensions)} \\ S &= \text{function(applied loads, densities, dimensions)} \end{aligned} \qquad (4.37)$$

Indeed R and S may not be statistically independent for example, cross-sectional dimensions affect both sectional strengths and dead loads. In this case, the best solution to the problem is to express each limit state equation or failure function in terms of the set of n *basic variables* $\overline{X}$ which affect the structural performance, such that

$$M = f(X_1, X_2, \ldots, X_n) \leqslant 0 \tag{4.38}$$

corresponds to failure. M is the *safety margin* sometimes referred to as the *failure indicator*. In general, the function f can take any form, provided that $M \leqslant 0$ corresponds to a failure state and $M > 0$ to a safe state. For simple problems, there is clearly no difficulty in finding a suitable form for f, but care should be taken in situations when some of the loads act in a resisting capacity (e.g. with loads resisting, as opposed to causing, overturning). In such cases, a check should be made that

$$\frac{\partial M}{\partial X_i} > 0 \tag{4.39}$$

where X_i is *any* resisting variable known to be active in the limit state under consideration. Setting

$$f(X_1, X_2, \ldots, X_n) = 0 \tag{4.40}$$

defines an $(n - 1)$ dimensional hyper-surface in the n-dimensional basic variable space. This surface is commonly referred to as the *failure surface* for the limit state under consideration and divides all possible combinations of the variables $\overline{X}$ which cause failure from all possible combinations which do not. It should be noted that this is an entirely deterministic concept. The reliability of the structure may then be expressed as

$$\mathcal{R} = 1 - P_f = 1 - \underbrace{\int\int\ldots\int}_{f(\overline{X}) \leqslant 0} f_{X_1, X_2, \ldots, X_n}(x_1, x_2, \ldots, x_n) dx_1\, dx_2 \ldots dx_n \tag{4.41}$$

where $f_{X_1, X_2, \ldots, X_n}(x_1, x_x, \ldots, x_n)$ is the joint probability density function for the n variables X_i. Note that the integral is over the failure region, denoted ω_f (see chapter 5.2).

If, and only if, the basic variables $\overline{X}$ are statistically independent, equation (4.41) may be replaced by

$$\mathcal{R} = 1 - P_f = 1 - \underbrace{\int\int\ldots\int}_{f(\overline{X}) \leqslant 0} f_{X_1}(x_1) f_{X_2}(x_2) \ldots f_{X_n}(x_n) dx_1\, dx_2 \ldots dx_n \tag{4.42}$$

Two practical problems are immediately apparent. First, there is almost never sufficient data to define the joint probability density function for the n basic variables. Usually, there is hardly enough information to be confident about the marginal distributions and the covariances. Secondly, even if the joint density function is knôwn, or in the case of equation (4.42) the marginal densities, the multi-dimensional integration required may be extremely time-consuming. Analytical solutions do not exist for the majority of practical problems and the analyst must resort to numerical methods.

These difficulties can be overcome in practice by using the level 2 methods described in chapters 5 and 6. The only other possibility is Monte-Carlo simulation, but this too has its limitations. However, for the sake of completeness, the technique will be briefly discribed.

4.3.6 Monte-Carlo Methods

Let us assume for simplicity that the basic variables X_i in equation (4.38) are statistically independent with known distribution functions. The Monte-Carlo approach is to use appropriate random number generators (see appendix A) to generate independent sample values x_i for each of the basic variables and to determine the corresponding value of the safety margin M from

$$m = f(x_1, x_2, \ldots, x_n) \tag{4.43}$$

By repeating this process many times it is possible to simulate the probability distribution for M by progressively building up a larger sample. This sample, although generated numerically, may be treated in the same way as any other statistical sample. In general, the exact probability distribution for M will not be of any standard form, although it may be governed by the form of the probability distribution of the most dominant basic variable.

The failure probability may be estimated in at least two ways. Since $M \leqslant 0$ corresponds to failure, P_f may be expressed as

$$P_f = P(M \leqslant 0) = \lim_{n \to \infty} k/n \tag{4.44}$$

where n is the total number of trials, and k is the number of trials in which $f(x_1, x_2, \ldots, x_n) \leqslant 0$.

However, the ratio k/n is a statistical variable whose sampling distribution, and in particular variance, depends on the number of trials n. For low failure probabilities and/or small n, the estimate of P_f given by k/n may be subject to considerable uncertainty. Practical rules can be established giving the necessary number of trials for any given magnitude of P_f.

The second approach is to fit an appropriate probability distribution $f_{M'}$, to the trial values of M', using its sample moments. Some suitable distributions are given by Elderton and Johnson [4.6]. See also [4.2]. This approach is the only one possible when the number of trials is small. Then

$$P_f \simeq \int_{-\infty}^{0} f_{M'}(m')dm' \tag{4.45}$$

As a general rule, however, Monte-Carlo methods should be avoided if at all possible.

BIBLIOGRAPHY

[4.1] Ang, A. H.-S. and Amin, M.: *Safety Factors and Probability in Structural Design.* Journal of the Structural Division, ASCE, July 1969, Vol. 95.

[4.2] Baker, M. J.: *The Reliability of Reinforced Concrete Floor Slabs in Office Buildings - A Probabilistic Study.* CIRIA Report 57, March 1976.

[4.3] Barlow, R. E. and Proschan, F.: *Mathematical Theory of Reliability.* John Wiley and Sons, 1965.

[4.4] Bolotin, V. V.: *Statistical Methods in Structural Mechanics.* Holden-Day, 1969.

[4.5] Cornell, C. A.: *Bayesian Statistical Decision Theory and Reliability-Based Design.* Int. Conf. Struc. Safety and Reliability of Eng. Struct., Washington, D.C., 9 - 11 April, 1969.

[4.6] Elderton, W. P. and Johnson, N. L.: *Systems of Frequency Curves.* Cambridge University Press, 1969.

[4.7] Ferry Borges, J. and Castanheta, M.: *Structural Safety.* National Laboratory of Civil Engineering, Lisbon, Portugal. 2nd Ed. 1971.

[4.8] Freudenthal, A. M.: *Safety, Reliability and Structural Design.* Transactions, ASCE, 1962, Vol. 127 (Part II).

[4.9] Freudenthal, A. M. and Gumbel, E. J.: *Failure and Survival in Fatigue.* Journal of Applied Physics, Vol. 25, No. 11, 1954.

[4.10] Freudenthal, A. M., Garretts, J. M. and Shinozuka, M.: *The Analysis of Structural Safety.* Journal of the Structural Division, ASCE, Vol. 92, No. ST1, Feb. 1966.

[4.11] Haugen, E. B.: *Probabilistic Approaches to Design.* John Wiley and Sons, 1968.

[4.12] Kapur, K. C. and Lamberson, L. R.: *Reliability Engineering Design.* John Wiley and Sons, 1977.

[4.13] Pugsley, A. G.: *The Safety of Structures.* Edward Arnold, 1966.

Chapter 5

LEVEL 2 METHODS

5.1 INTRODUCTION

In chapters 1 and 4 an introduction to the fundamental concepts of structural reliability is given and the so-called level 2 methods are briefly mentioned. In this chapter, level 2 methods will be treated in a more detailed way and a number of simple applications will be shown. In chapter 6 further applications of level 2 methods with special emphasis on non-normally distributed basic variables and correlated basic variables are treated.

As mentioned in section 1.3.2, methods of reliability analysis are classified on the basis of the types of calculations performed and of the approximations made. The most advanced methods are the *level 3 methods.* They can be characterized as being probabilistic methods of analysis based on knowledge of the (joint) distribution of all basic variables. In *level 2 methods* a number of idealizations compared with the »exact» level 3 methods are made. In particular it is assumed that the failure surface described in 5.2 below can be sensibly approximated by a tangent hyperplane at the point on the failure surface closest to the origin, when the surface has been mapped into a standard normal space. A level 2 method is therefore a method of design or analysis which in its simplest form comprises a check at only a single point on the failure surface, as opposed to level 3 methods where the probability content of the entire failure region is evaluated.

Level 2 methods provide a powerful set of tools for tackling a wide range of practical problems. The relatively simple structural examples given in this chapter and in chapter 6 are for the purpose of explaining the methods. More complex and practical applications are discussed in chapters 11 and 12.

5.2 BASIC VARIABLES AND FAILURE SURFACES

One of the first problems one has to solve before the reliability of a given structure can be evaluated is to decide upon which variables (quantities or parameters) are of relevance. These variables called *basic variables* can be geometrical quantities (e.g. the area of the cross-section of a beam), material strength (e.g. rupture or yield strength) and external loads (e.g. traffic loads, wave or wind loads). For a given structure each variable x_i, $i = 1, \ldots, n$ is considered a realization of a random variable X_i, $i = 1, \ldots, n$ and therefore, the set of variables $\bar{x} = (x_1, \ldots, x_n)$ is a realization of the random vector $\bar{X} = (X_1, \ldots, X_n)$. In other words, the variable $\bar{x}$ is a point in an n-dimensional *basic variable space.*

In a level 3 method knowledge of the joint probability density function $f_{\overline{X}}$ is required, but in the level 2 method presented in this chapter only the expectations

$$\mu_i = E[X_i] \quad , \quad i = 1, \ldots, n \tag{5.1}$$

and the covariances

$$C_{ij} = \mathrm{Cov}[X_i, X_j] \quad , \quad i, j = 1, 2, \ldots, n \tag{5.2}$$

are used. Note that the variance $\mathrm{Var}[X_i] = \sigma_i^2$ is equal to $\mathrm{Cov}[X_i, X_i]$.

In the following it is assumed that a set of basic variables $\overline{X} = (X_1, \ldots, X_n)$ is chosen in such a way that a *failure surface* (or *limit state surface*) can be defined in the n-dimensional basic variable space ω. A failure surface is a surface dividing the basic variable space into two regions namely a *failure region* ω_f and a *safe region* ω_s. The failure region contains all realizations of $\overline{X}$ that would result in failure, and the safe region contains all realizations of X that would not result in failure. It is convenient to describe the failure surface by an equation of the form

$$f(\overline{x}) = f(x_1, x_2, \ldots, x_n) = 0 \tag{5.3}$$

in such a way that positive values of f indicate safe sets of basic variables (the safe region) and non-positive values of f indicate unsafe sets of variables (the failure region), i.e.

$$f(\overline{x}) = \begin{cases} > 0 & \text{when } \overline{x} \in \omega_s \\ \leq 0 & \text{when } \overline{x} \in \omega_f \end{cases} \tag{5.4}$$

A 2-dimensional case is illustrated in figure 5.1. The function $f: \omega \to R$ is called the *failure function.* It is important to note that the same failure surface can be described by a number of equivalent failure functions.

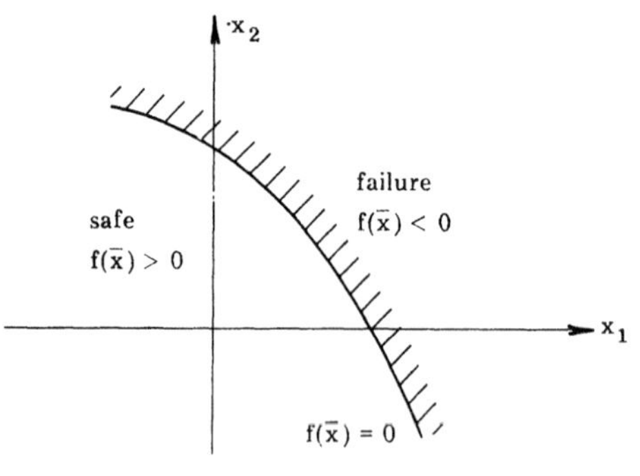

Figure 5.1.

Let f be a failure function. The random variable $M = f(\overline{X})$ is then called a *safety margin.*

Example 5.1. Consider the fundamental case with only two basic variables (a load variable S and a strength parameter R) and a failure function $f_1 : R^2 \to R$, where

$$f_1(r, s) = r - s \tag{5.5}$$

The failure surface, the failure region and the safe region are shown in figure 5.2. The corresponding safety margin M_1 is given by

$$M_1 = R - S \tag{5.6}$$

An equivalent failure function is $f_2 : R^2 \to R$, where

$$f_2(r, s) = \ell n \frac{r}{s} = \ell n\, r - \ell n\, s \tag{5.7}$$

with the safety margin

$$M_2 = \ell n \frac{R}{S} = \ell n\, R - \ell n\, S \tag{5.8}$$

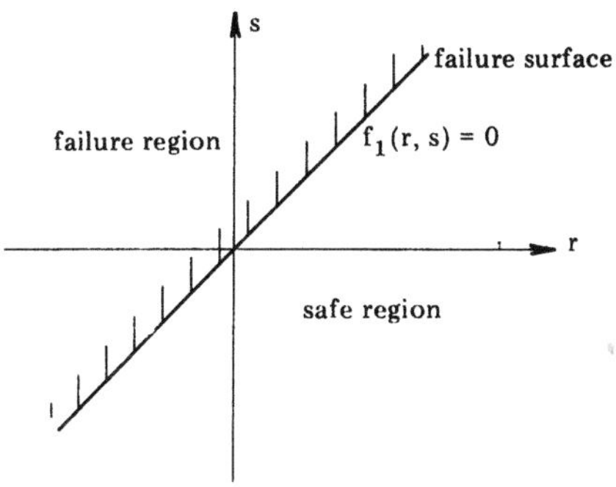

Figure 5.2

5.3 RELIABILITY INDEX FOR LINEAR FAILURE FUNCTIONS AND NORMAL BASIC VARIABLES

For the case of a linear safety margin M and normal basic variables, the reliability index β is defined by

$$\beta = \frac{\mu_M}{\sigma_M} \tag{5.9}$$

where μ_M is the mean of M and σ_M is the standard deviation of M. This definition of the reliability index β was used by Cornell [5.1] as early as 1969.

Example 5.2. Consider the fundamental case treated in example 5.1 and assume that R and S are uncorrelated. With M = R − S one gets

$$\mu_M = \mu_R - \mu_S \tag{5.10}$$

and

$$\sigma_M^2 = \sigma_R^2 + \sigma_S^2 \tag{5.11}$$

according to equation (2.86). Therefore

$$\beta = \frac{\mu_R - \mu_S}{(\sigma_R^2 + \sigma_S^2)^{\frac{1}{2}}} \tag{5.12}$$

Let the safety margin M be *linear* in the basic variables $X_1, \ldots, X_n$

$$M = a_0 + a_1 X_1 + \ldots + a_n X_n \tag{5.13}$$

It is then easy to calculate the reliability index β as

$$\mu_M = a_0 + a_1 \mu_1 + \ldots + a_n \mu_n \tag{5.14}$$

and

$$\sigma_M^2 = a_1^2 \sigma_1^2 + \ldots + a_n^2 \sigma_n^2 + \sum_{i=1}^{n} \sum_{j=1,\, j \neq i}^{n} \rho_{ij} a_i a_j \sigma_i \sigma_j \tag{5.15}$$

where the last term accounts for correlation between any pair of basic variables. ρ_{ij} is equal to the correlation coefficient $\rho_{X_i X_j}$ defined in equation (2.80).

Example 5.3. Consider the statically indeterminate beam shown in figure 5.3 loaded by a concentrated force p and assume that the beam fails when $|m| \geqslant m_F$, where m_F is a critical limit moment and m is the maximum moment in the beam. Further assume that p, ℓ and m_F are realizations of uncorrelated random variables P, L, and M_F with

$\mu_P = 4$ kN , $\sigma_P = 1$ kN

$\mu_L = 5$ m , $\sigma_L \sim 0$ m

$\mu_{M_F} = 20$ kNm , $\sigma_{M_F} = 2$ kNm

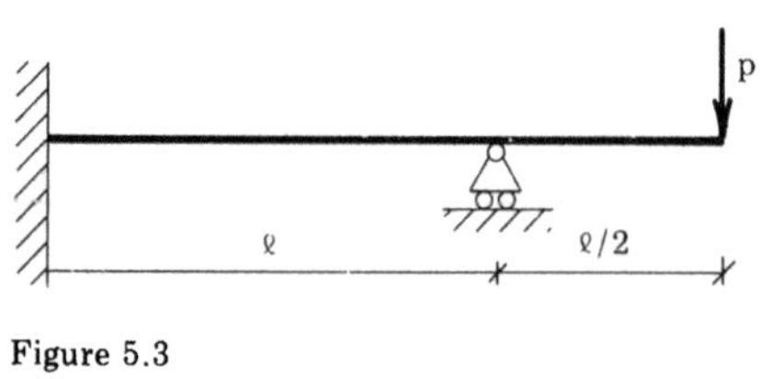

Figure 5.3

: The maximum moment is $|m| = p\ell/2$ and therefore, the following failure function can
: be used

$$f(p, \ell, m_F) = m_F - \frac{1}{2} p\ell = 0 \tag{5.16}$$

: Note that ℓ in this case can be considered a deterministic parameter because $\sigma_L \sim 0$ m.
: Therefore (5.16) can be rewritten

$$f(p, m_F) = m_F - \frac{5}{2} p = 0$$

: The corresponding safety margin is

$$M(P, M_F) = M_F - \frac{5}{2} P \tag{5.17}$$

: and

$$\mu_M = 20 - \frac{5}{2} \cdot 4 = 10 \text{ kNm}$$
$$\sigma_M^2 = 4 + \frac{25}{4} \cdot 1 = 10.25 \text{ (kNm)}^2 \tag{5.18}$$

: Therefore

$$\beta = \frac{10}{\sqrt{10.25}} = 3.12 \tag{5.19}$$

Note that in the presentation above, a safety margin linear in the basic variables has been assumed. If the safety margin M is *non-linear* in $\overline{X} = (X_1, \ldots, X_n)$ then approximate values for μ_M and σ_M can be obtained by using a linearized safety margin M. Let

$$M = f(\overline{X}) = f(X_1, \ldots, X_n) \tag{5.20}$$

By expanding this relationship in a Taylor series about $(X_1, \ldots, X_n) = (\mu_1, \ldots, \mu_n)$ and retaining only the linear terms one gets

$$M = f(\overline{X}) \cong f(\mu_1, \ldots, \mu_n) + \sum_{i=1}^{n} \frac{\partial f}{\partial X_i} (X_i - \mu_i) \tag{5.21}$$

where $\partial f / \partial X_i$ is evaluated at $(\mu_1, \ldots, \mu_n)$. From (5.21) approximate values for μ_M and σ_M are determined by

$$\mu_M \cong f(\mu_1, \ldots, \mu_n) \tag{5.22}$$

$$\sigma_M^2 \cong \sum_{i=1}^{n} \sum_{j=1}^{n} \frac{\partial f}{\partial X_i} \frac{\partial f}{\partial X_j} \operatorname{Cov}[X_i, X_j] \tag{5.23}$$

: **Exercise 5.1.** Prove equations (5.22) and (5.23) and derive the special form of (5.23) when the basic variables are uncorrelated.

Clearly, for non-linear failure functions, calculation of the reliability index $\beta = \mu_M/\sigma_M$ on the basis of a linearization as (5.21) will depend on the choice of linearization point. In (5.21) the so-called *mean point* $(\mu_1, \ldots, \mu_n)$ is chosen, but as shown later, a point on the failure surface would be more reasonable. Experience shows that an expansion based on the mean point should not be used.

The reliability index β as defined by equation (5.9) will change when different but equivalent non-linear failure functions are used. This can easily be demonstrated by considering the fundamental case treated in example 5.2. In the fundamental case only two uncorrelated basic variables R and S are involved. As shown in example 5.2 the safety margin M = R − S results in the reliability index

$$\beta = \frac{\mu_R - \mu_S}{(\sigma_R^2 + \sigma_S^2)^{\frac{1}{2}}} \tag{5.24}$$

By using the equivalent safety margin (5.8) $M = \ell n(R/S) = \ell n R - \ell n S$ one gets

$$\beta' = \frac{\mu_{\ell n(R/S)}}{\sigma_{\ell n(R/S)}} \tag{5.25}$$

: **Exercise 5.2.** Show that the following approximate value of β' defined by (5.25)

$$\beta' \cong \frac{\ell n \mu_R - \ell n \mu_S}{\sqrt{(\frac{\sigma_R}{\mu_R})^2 + (\frac{\sigma_S}{\mu_S})^2}} \tag{5.26}$$

: is obtained by linearization of the safety margin $M = \ell n(R/S)$ about (μ_R, μ_S).

The values β and β' from (5.24) and (5.26) are different. The reliability index defined by equation (5.9) is thus not invariant with regard to the choice of failure function. It is of course unfortunate that a reliability measure can give different values for the same problem. In the next section the reliability index is redefined, so that this problem is solved in a simple way.

In general, it is not possible to relate the reliability index β defined by (5.9) to the probability of failure P_f given by

$$P_f = \int_{\omega_f} f_{\bar{X}}(\bar{x}) d\bar{x} \tag{5.27}$$

where $f_{\bar{X}}$ is the joint probability density function and ω_f is the failure region defined earlier in this section. However, when the safety margin M is linear in the basic variables X_i, i = 1, ... , n, and these basic variables are normally distributed $N(\mu_i, \sigma_i)$, then the following relationship exists,

$$P_f = \Phi(-\beta) \Longleftrightarrow \beta = -\Phi^{-1}(P_f) \tag{5.28}$$

where Φ is the standardized normal distribution function. Here, (5.28) will only be shown in the 2-dimensional fundamental case with 2 independent basic variables R and S, and with the safety margin $M = R - S$, where R is normally distributed $N(\mu_R, \sigma_R)$ and S is normally distributed $N(\mu_S, \sigma_S)$. Then, M is normally distributed $N(\mu_R - \mu_S, \sqrt{\sigma_R^2 + \sigma_S^2})$, and the probability of failure is

$$P_f = P(R - S \leqslant 0) = P(M \leqslant 0) = \Phi\left(\frac{0 - (\mu_R - \mu_S)}{\sqrt{\sigma_R^2 + \sigma_S^2}}\right) = \Phi(-\beta) \tag{5.29}$$

according to (5.24) (see also Section 4.3.2).

Example 5.4. Consider again the beam in example 5.3, where the reliability index β = 3.12. If the basic variables M_F and P are normally distributed the probability of failure can be calculated by (5.29)

$$P_f = \Phi(-\beta) = \Phi(-3.12) = 0.0009$$

Consider again the 2-dimensional fundamental case with independent basic variables R and S and let the means be μ_R and μ_S and the standard deviations σ_R and σ_S. Further, let the safety margin be $M = R - S$. The reliability index β can then be given a simple geometrical interpretation in a normalized coordinate system as shown in figure 5.4, where the coordinates r' and s' are realizations of random variables $R' = (R - \mu_R)/\sigma_R$ and $S' = (S - \mu_S)/\sigma_S$. With these variables, the failure surface is given by

$$\sigma_R R' - \sigma_S S' + (\mu_R - \mu_S) = 0 \tag{5.30}$$

Therefore, the shortest distance from the origin to this linear failure surface is equal to the reliability index β (see figure 5.4 and (5.24)).

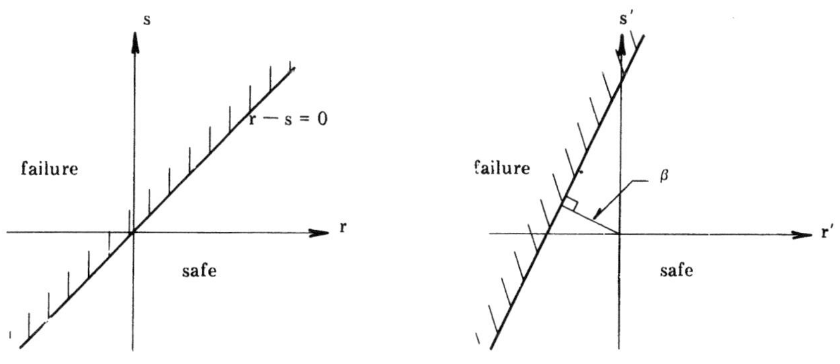

Figure 5.4

5.4 HASOFER AND LIND'S RELIABILITY INDEX

A serious objection to the reliability index β as defined by equation (5.9) is its lack of failure function invariance as discussed in section 5.3. In this section the reliability index β proposed by Hasofer and Lind in 1974 will be introduced, and it will be shown that it is invariant with respect to the choice of failure function.

As before, the basic variables are called $\overline{X} = (X_1, X_2, \ldots, X_n)$ and the failure function $f : \omega \to R$, where ω is the n-dimensional basic variable space. The space is divided into a failure region $\omega_f = \{\overline{x}: f(\overline{x}) \leq 0\}$, and a safe region $\omega_s = \{\overline{x}: f(\overline{x}) > 0\}$ by the failure surface $\partial\omega = \{\overline{x}: f(\overline{x}) = 0\}$.

In this section it is assumed for the sake of simplicity that the basic variables are *uncorrelated*, i.e. $\mathrm{Cov}[X_i, X_j] = 0$ for all i and j. It will be shown in the next chapter how to deal with safety problems when this assumption is relaxed.

The first step in defining Hasofer and Lind's reliability index is to normalize the set of basic variables. This new set $\overline{Z} = (Z_1, \ldots, Z_n)$ is defined by

$$Z_i = \frac{X_i - \mu_{X_i}}{\sigma_{X_i}} \quad , \quad i = 1, 2, \ldots, n \tag{5.31}$$

where μ_{X_i} and σ_{X_i} are the mean and the standard deviation of the random variable X_i. Note that

$$\mu_{Z_i} = 0 \quad \text{and} \quad \sigma_{Z_i} = 1 \quad , \quad i = 1, 2, \ldots, n \tag{5.32}$$

By the linear mapping defined by (5.31) the failure surface in the x-coordinate system is mapped in to a failure surface in the z-coordinate system. The failure surface in the z-coordinate system divides the z-space into a failure region and a safe region in the same way as in the x-space. Due to equation (5.32) the new z-coordinate system has an important characteristic, namely a rotational symmetry with respect to the standard deviations. Note that the origin O of the normalized z-coordinate system will usually be within the safe region. A two-dimensional example is shown in figure 5.5.

Hasofer and Lind's reliability index β is defined as the shortest distance from the origin to the failure surface in the normalized z-coordinate system. For the two-dimensional case in figure 5.5 β is equal to the distance OA. The point A is called the *design point.* By using a definition of the reliability index β, where β is related to the failure surface and not to the failure function, a safety measure is obtained which is failure function invariant because all equivalent failure functions result in the same failure surface.

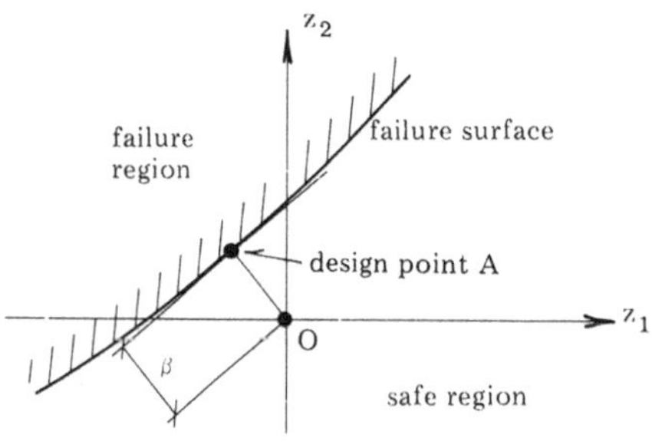

Figure 5.5

The reliability index β as defined by equation (5.9) will coincide with the reliability index β defined above, when the failure surface is linear (a hyperplane). This is shown for the two-dimensional case at the end of last section, but can easily be proven in the general n-dimensional case. Therefore, in this case the important relation (5.28) between β and the probability of failure P_f can also be formulated, provided the basic variables X_i, i=1, 2, ...,n are normally distributed

$$P_f = \Phi(-\beta) \Leftrightarrow \beta = -\Phi^{-1}(P_f) \tag{5.33}$$

In the previous section it was shown how an approximate calculation of the reliability index β defined by eq. (5.9) can be obtained when the safety margin M is non-linear by expanding M in a Taylor Series about the point $(X_1, \ldots, X_n) = (\mu_1, \ldots, \mu_n)$. It is now obvious from the above remarks that the two definitions of β will coincide if this expansion is made about the design point. This corresponds to approximating the non-linear failure surface by its tangent plane at the design point (A in figure 5.5).

The definition of the reliability index β by Hasofer and Lind can be formulated in the following way

$$\beta = \min_{\bar{z} \in \partial\omega} \left(\sum_{i=1}^{n} z_i^2\right)^{\frac{1}{2}} \tag{5.34}$$

where $\partial\omega$ is the failure surface in the z-coordinate system. The calculation of β can be undertaken in a number of different ways (see also page 227). In the general case where the failure surface is non-linear an iterative method must be used. Here an iteration method with fast rate of convergence is given. The procedure will be illustrated by a number of simple examples where the failure function is assumed to be differentiable. In this case the distance β and the unit vector $\bar{\alpha} = (\alpha_1, \ldots, \alpha_n)$ given by $\overrightarrow{OA} = \beta\bar{\alpha}$, where A is the design point, can be determined by solving the following n + 1 equations iteratively

$$\left.\begin{aligned} \alpha_i &= \frac{-\dfrac{\partial f}{\partial z_i}(\beta\bar{\alpha})}{\left[\sum_{k=1}^{n}\left(\dfrac{\partial f}{\partial z_k}(\beta\bar{\alpha})\right)^2\right]^{\frac{1}{2}}}, \quad i = 1, 2, \ldots, n \\ f(\beta\alpha_1, \beta\alpha_2, &\ldots, \beta\alpha_n) = 0 \end{aligned}\right\} \tag{5.35}$$

where the failure surface is given by

$$f(z_1, z_2, \ldots, z_n) = 0 \tag{5.36}$$

Example 5.5. Consider the same beam as in example 5.3 but now with the following deflection failure criterion

$$u_{max} = \frac{5}{48}\frac{p\ell^3}{ei} \geqslant \frac{1}{30}\ell \tag{5.37}$$

where u_{max} is the maximum deflection, e the modulus of elasticity and i the relevant moment of inertia. Further, let u_{max}, p, ℓ, e and i be realizations of uncorrelated random variables U_{max}, P, L, E and I with

$\mu_P = 4$ kN ; $\sigma_P = 1$ kN

$\mu_L = 5$ m ; $\sigma_L \cong 0$ m

$\mu_E = 2 \cdot 10^7$ kN/m² ; $\sigma_E = 0.5 \cdot 10^7$ kN/m²

$\mu_I = 10^{-4}$ m⁴ ; $\sigma_I = 0.2 \cdot 10^{-4}$ m⁴

By inserting $\ell = 5$ m into (5.37) the failure function can be written

$$ei - 78.12 \cdot p = 0 \tag{5.38}$$

The basic variables I, E and P are then normalized $Z_1 = (I - \mu_I)/\sigma_I$, $Z_2 = (E - \mu_E)/\sigma_E$ and $Z_3 = (P - \mu_P)/\sigma_P$. In the normalized coordinate system the failure surface is given by

$$(10^{-4} + 0.2 \cdot 10^{-4} z_1) \cdot (2 \cdot 10^7 + 0.5 \cdot 10^7 z_2) - 78.12 \cdot (4 + z_3) = 0$$

or

$$0.2 z_1 + 0.25 z_2 + 0.05 z_1 z_2 - 0.0391 z_3 + 0.8438 = 0$$

The reliability index β and the design point are then determined by the following equations

$$\left.\begin{aligned} \beta &= \frac{-0.8438}{0.2\alpha_1 + 0.25\alpha_2 + 0.05\beta\alpha_1\alpha_2 - 0.0391\alpha_3} \\ \alpha_1 &= -\frac{1}{k}(0.2 + 0.05\,\beta\alpha_2) \\ \alpha_2 &= -\frac{1}{k}(0.25 + 0.05\,\beta\alpha_1) \\ \alpha_3 &= \frac{1}{k}0.0391 \end{aligned}\right\} \tag{5.39}$$

where the same notation as in (5.35) is used and where k is determined by putting

$$\sum_{i=1}^{3} \alpha_i^2 = \alpha_1^2 + \alpha_2^2 + \alpha_3^2 = 1$$

The first equation in (5.39) is written in a form suitable for iteration. The iteration is now performed by choosing starting values for $\beta_1 \alpha_1$, α_2, and α_3 and calculating new values by (5.39) until only small modifications are obtained. This is shown in table 5.1.

	Start	Iteration No.				
		1	2	3	4	5
β	3	3.62	3.51	3.36	3.30	3.29
α_1	−0.58	−0.53	−0.33	−0.23	−0.19	−0.18
α_2	−0.58	−0.82	−0.91	−0.95	−0.97	−0.97
α_3	+ 0.58	0.22	0.23	0.19	0.18	0.17

Table 5.1

The reliability index is $\beta = 3.29$ and the design point A = 3.29(− 0.18, − 0.97, 0.17) = (− 0.59, − 3.19, 0.56). Note that the correct sign for α_i, i = 1, 2, 3, was chosen for the starting values in the table. This can usually be done in the following way. A positive sign can be recommended, when the corresponding basic variable is a »loading variable» (e.g. P) and a negative sign when it is a »strength or geometrical variable» (e.g. E and I).

Exercise 5.3. Consider the elastic beam shown on figure 5.6 with a uniform load p, length ℓ and critical limit moment m_F. Assume that p, ℓ and m_F are realizations of uncorrelated random variables P, L and M_F with

μ_P = 2 Mp/m σ_P = 0.4 Mp/m

μ_L = 4 m σ_L = 0.4 m

μ_{M_F} = 5 Mpm σ_{M_F} = 0.4 Mpm

The maximum bending moment is $m_{max} = \frac{9}{128} p\ell^2$. Calculate the reliability index β for the following failure mode $m_{max} \geq m_F$.

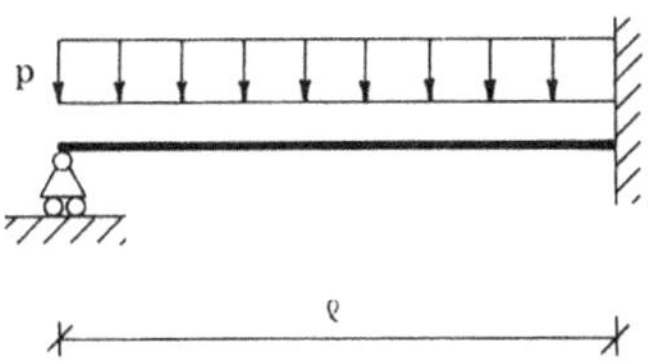

Figure 5.6

In examples 5.3 and 5.5 and exercise 5.3, structures with given loading, strength and dimensions have been *analysed*. The reliability index is calculated by an iteration method suitable for hand calculations and computer calculations. It will be shown in example 5.6 that the same techniques can be used in *designing* a structure so that it achieves a given reliability index β.

Example 5.6. Consider the elastic beam shown on figure 5.7 loaded with a single load p. The maximum deflection is $u_{max} = \frac{1}{192} \frac{p\ell^3}{ei}$, where ℓ is the length of the beam. Assume that p, ℓ, e and i are realizations of uncorrelated random variables P, L, E and I, with

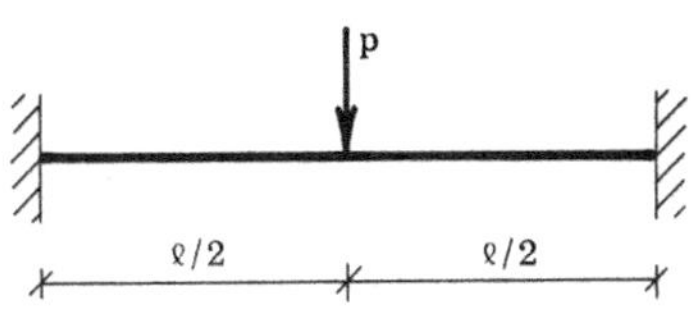

Figure 5.7

$\mu_P = 4$ kN $\qquad \sigma_P = 1$ kN

$\mu_L = 6$ m $\qquad \sigma_L \approx 0$ m

$\mu_E = 2 \cdot 10^7$ kN/m² $\qquad \sigma_E = 0.5 \cdot 10^7$ kN/m²

The mean value μ_I for the moment of inertia is unknown and the problem here is to determine μ_I so that $\beta = 3.0$, when the failure mode is $u_{max} \geqslant \frac{1}{100} \cdot \ell$ and when $\sigma_I = 0.1 \cdot \mu_I$. The failure function is defined by

$$\frac{1}{100}\ell - \frac{1}{192}\frac{p\ell^3}{ei} = 0$$

or

$$6ei - 113\,p = 0$$

The basic variables P, E and I are normalized $Z_1 = (P - \mu_P)/\sigma_P$, $Z_2 = (E - \mu_E)/\sigma_E$ and $Z_3 = (I - \mu_I)/\sigma_I$. In the normalized coordinate system the failure function is given by

$$6 \cdot 10^7 (2 + 0.5\, z_2)\mu_I(1 + 0.1\, z_3) - 113(4 + z_1) = 0$$

The design point is now given by $z_i = \beta\alpha_i = 3\alpha_i$. The unknown μ_I, α_1, α_2 and α_3 are therefore determined by the following system of equations:

$$\left.\begin{aligned} \mu_I &= \frac{113(4 + 3\alpha_1)}{6 \cdot 10^7 (2 + 1.5\,\alpha_2)(1 + 0.3\,\alpha_3)} \\ \alpha_1 &= \frac{1}{k}\,113 \\ \alpha_2 &= -\frac{1}{k} \cdot 10^7 (3 + 0.9\,\alpha_3)\mu_I \\ \alpha_3 &= -\frac{1}{k} \cdot 10^7 (1.2 + 0.9\,\alpha_2)\mu_I \end{aligned}\right\} \qquad (5.40)$$

where k is determined by the condition $\alpha_1^2 + \alpha_2^2 + \alpha_3^2 = 1$. This system can be solved iteratively in the same way as the system (5.39) in example 5.5 was solved.

	Start	Iteration No.			
		1	2	3	4
μ_I	$40 \cdot 10^{-7}$	$116 \cdot 10^{-7}$	$158 \cdot 10^{-7}$	$167 \cdot 10^{-7}$	$167 \cdot 10^{-7}$
α_1	0.58	0.35	0.23	0.23	0.23
α_2	−0.58	−0.90	−0.96	−0.97	−0.97
α_3	−0.58	−0.25	−0.14	−0.11	−0.11

Table 5.2

The mean value for the moment of inertia corresponding to $\beta = 3$ is $\mu_I = 167 \cdot 10^{-7}$ m⁴.

: **Exercise 5.4.** Consider the same elastic beam as in exercise 5.3, but with μ_{M_F} assumed un-
: known. Determine μ_{M_F} so that $\beta = 2.9$ with the same failure function and with $\sigma_{M_F} = 0.08\mu_{M_F}$.

BIBLIOGRAPHY

[5.1] Cornell, C. A.: *A Probability-Based Structural Code.* ACI-Journ., Vol. 66, 1969, pp. 974-985.

[5.2] Ditlevsen, O.: *Structural Reliability and the Invariance Problem.* Research Report No. 22, Solid Mechanics Division, University of Waterloo, Canada, 1973.

[5.3] Ditlevsen, O.: *Fundamentals of Second Moment Structural Reliability Theory.* Int. Research Seminar on Safety of Structures, Trondheim, Norway, 1977.

[5.4] Dyrbye, C. et al.: *Konstruktioners sikkerhed* (in Danish). Den Private Ingeniørfond ved Danmarks tekniske Højskole, København, 1979.

[5.5] Gravesen, S.: *Level II Safety Methods.* Lectures on Structural Reliability (ed. P. Thoft-Christensen), Aalborg University Centre, Aalborg, Denmark, 1980, pp. 29-38.

[5.6] Hasofer, A. M. and N. C. Lind: *An Exact and Invariant First Order Reliability Format.* Proc. ASCE, J. Eng. Mech. Div., 1974, pp. 111 - 121.

[5.7] Thoft-Christensen, P.: *Some Experience from Application of Optimization Technique in Structural Reliability.* Bygningsstatiske Meddelelser, Vol. 48, 1977, pp. 31-44.

Chapter 6

EXTENDED LEVEL 2 METHODS

6.1 INTRODUCTION

In chapter 5 a detailed presentation of level 2 methods was given. It was shown that the reliability index β of Hasofer and Lind is failure function invariant in the sense that equivalent failure functions result in the same reliability index.

Calculation of the reliability index has been shown in chapter 5 by a number of examples, but only *uncorrelated basic variables* have been treated. In this chapter the treatment will be extended so that *correlated basic variables* can be included.

This extension is important because in practical applications basic variables will often be correlated and because correlation sometimes affects the value of β appreciably.

In section 5.4 it was shown that the probability of failure P_f can be related to the reliability index β, when the following two conditions are fulfilled

(a) the failure surface is linear

(b) the basic variables X_i, i = 1, 2, . . . , n are normally distributed.

The one-one relation between P_f and β is given by equation (5.33)

$$P_f = \Phi(-\beta) \iff \beta = -\Phi^{-1}(P_f) \tag{5.33}$$

When the basic variables are non-normally distributed this one-to-one relation is not valid. To retain this relation for non-normally distributed basic variables also, it is natural to approximate a non-normally distributed variable with a normally distributed variable. In this chapter it will be shown how this can be undertaken.

6.2 CONCEPT OF CORRELATION

The concept of correlation has already been introduced in section 2.9. Let X_1 and X_2 be two random variables with the expected values $E[X_1] = \mu_{X_1}$ and $E[X_2] = \mu_{X_2}$. Then the *covariance* of X_1 and X_2 is denoted $Cov[X_1, X_2]$ and is defined by (see page 33)

$$Cov[X_1, X_2] = E[(X_1 - \mu_{X_1})(X_2 - \mu_{X_2})] = E[X_1 \cdot X_2] - E[X_1]\ E[X_2] \tag{6.1}$$

The ratio

$$\rho_{X_1 X_2} = \frac{Cov[X_1, X_2]}{\sigma_{X_1} \cdot \sigma_{X_2}} \tag{6.2}$$

where σ_{X_1} and σ_{X_2} are the standard deviations of X_1 and X_2 is called the *correlation coefficient*. It can be shown that

$$-1 \leq \rho_{X_1 X_2} \leq 1 \tag{6.3}$$

The random variables are said to be *uncorrelated* if $Cov[X_1, X_2] = 0$.

It follows from (6.1) that if a large and positive value of $Cov[X_1, X_2]$ occurs then the values of X_1 and X_2 tend to be simultaneously large or small relative to their means. If the value of $Cov[X_1, X_2]$ is numerically large but negative, then the values of X_1 tend to be large when the the values of X_2 are small relative to their means, and vice versa. When $Cov[X_1, X_2]$ is close to zero there is no linear relationship between X_1 and X_2.

Example 6.1. Consider the beam AB shown in figure 6.1 loaded with two random loads P_1 and P_2. The loads are assumed statistically independent with means $\mu_{P_1} = 4$ kN and $\mu_{P_2} = 6$ kN and standard deviations $\sigma_{P_1} = 0.4$ kN and $\sigma_{P_2} = 0.5$ kN. The shear force Q_B and bending moment M_B at the support B are

$$Q_B = \frac{1}{27}(13\,P_1 + 23\,P_2)$$

$$M_B = \frac{6}{27}(4\,P_1 + 5\,P_2)$$

Q_B and M_B are random variables with the means

$$\mu_{Q_B} = \frac{1}{27}(13\,\mu_{P_1} + 23\,\mu_{P_2}) = \frac{190}{27}\ \text{kN}$$

$$\mu_{M_B} = \frac{6}{27}(4\,\mu_{P_1} + 5\,\mu_{P_2}) = \frac{92}{9}\ \text{kNm}$$

and the variances

$$\sigma_{Q_B}^2 = \frac{1}{27^2}(13^2\sigma_{P_1}^2 + 23^2\sigma_{P_2}^2) = 0.2185 \quad , \quad \sigma_{Q_B} = 0.467\ \text{kN}$$

$$\sigma_{M_B}^2 = \frac{6^2}{27^2}(4^2\sigma_{P_1}^2 + 5^2\sigma_{P_2}^2) = 0.4351 \quad , \quad \sigma_{M_B} = 0.660\ \text{kNm}$$

The loads P_1 and P_2 are statistically independent, but Q_B and M_B are correlated. The covariance $Cov[Q_B M_B]$ can be evaluated by (6.1), where

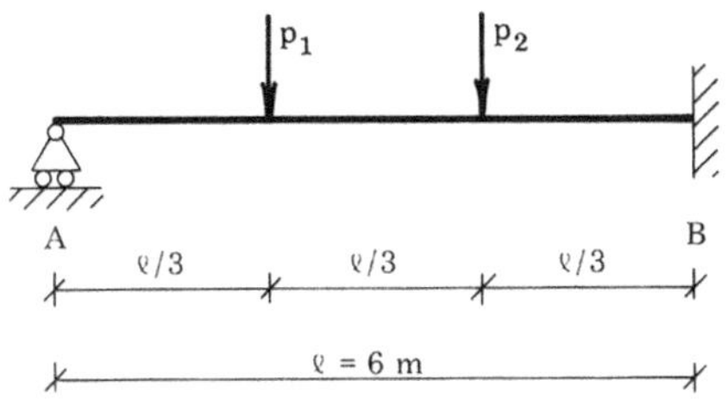

Figure 6.1.

$$E[Q_B M_B] = E[\frac{1}{27}(13\,P_1 + 23\,P_2)\frac{6}{27}(4\,P_1 + 5\,P_2)]$$

$$= \frac{6}{27^2}(52\,E[P_1^2] + 157\,E[P_1\,P_2] + 115\,E[P_2^2])$$

In this equation

$$E[P_1\,P_2] = E[P_1]E[P_2] = 24(kN)^2$$

$$E[P_1^2] = \sigma_{P_1}^2 + \mu_{P_1}^2 = 0.16 + 16 = 16.16\,(kN)^2$$

$$E[P_2^2] = \sigma_{P_2}^2 + \mu_{P_2}^2 = 0.25 + 36 = 36.25\,(kN)^2$$

Therefore

$$E[Q_B M_B] = \frac{6}{27^2}(52\cdot 16.16 + 157\cdot 24 + 115\cdot 36.25) = 72.239\,(kN)^2\,m$$

and

$$Cov[Q_B M_B] = E[Q_B M_B] - \mu_{Q_B}\mu_{M_B} = 72.239 - \frac{190}{27}\cdot\frac{92}{9} = 0.3051\,(kN)^2\,m$$

Finally, the correlation coefficient is

$$\rho_{Q_B M_B} = \frac{0.3051}{0.467\cdot 0.660} = 0.99$$

This value close to one indicates that Q_B and M_B are strongly correlated.

Exercise 6.1. Consider the cantilever beam AB shown in figure 6.2, loaded with two statistically independent random loads P_1 and P_2, where $\mu_{P_1} = \mu_{P_2} = 5$ kN and $\sigma_{P_1} = \sigma_{P_2} = 1$ kN. Let Q_B and M_B be the shear force and bending moment at the support B. Show that

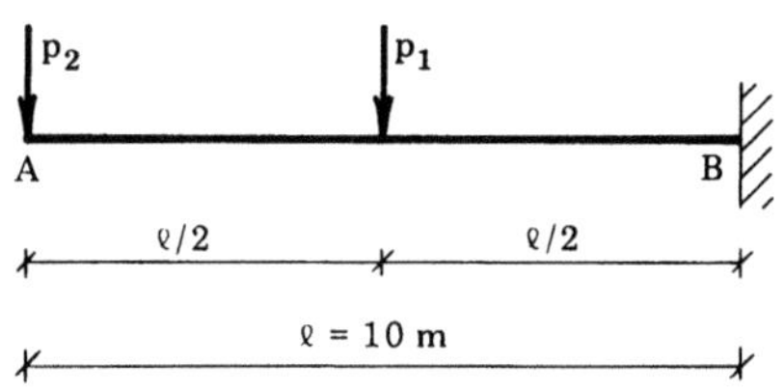

Figure 6.2

: $\mu_{Q_B} = 10$ kN $\sigma_{Q_B} = 1.4142$ kN
:
: $\mu_{M_B} = 75$ kNm $\sigma_{M_B} = 11.1803$ kNm
:
: and
:
: $\rho_{Q_B M_B} = 0.95$

In example 6.1 and exercise 6.1 random variables Q_B and M_B both depending on the same variables P_1 and P_2 are considered. Therefore, Q_B and M_B are correlated although P_1 and P_2 are uncorrelated.

Very often one needs to calculate the mean and variance for a random variable Y which is a linear function of a number of random variables $X_1, \ldots, X_n$

$$Y = a_0 + a_1 X_1 + \ldots + a_n X_n \tag{6.3}$$

The expected value E[Y] is then given by

$$E[Y] = a_0 + \sum_{i=1}^{n} a_i E[X_i] \tag{6.4}$$

and the variance Var[Y] by

$$\mathrm{Var}[Y] = \sum_{i=1}^{n} a_i^2 \,\mathrm{Var}[X_i] + \sum_{i \neq j}^{n} \sum^{n} a_i a_j \,\mathrm{Cov}[X_i, X_j] \tag{6.5}$$

where equation (6.5) is a generalization of equation (2.86), see also (5.15).

Let the random variables Y_1 and Y_2 be linear functions of $X_1, X_2, \ldots, X_n$

$$Y_1 = \sum_{i=1}^{n} a_i X_i \tag{6.6}$$

$$Y_2 = \sum_{i=1}^{n} b_i X_i \tag{6.7}$$

It can then be shown that

$$\mathrm{Cov}[Y_1, Y_2] = \sum_{i=1}^{n} a_i b_i \,\mathrm{Var}[X_i] + \sum_{i \neq j}^{n} \sum^{n} a_i b_j \,\mathrm{Cov}[X_i, X_j] \tag{6.8}$$

: **Example 6.2.** Consider again the problem solved in example 6.1. The covariance between
: Q_B and M_B can now more easily be calculated by equation (6.8)
:
: $\mathrm{Cov}[Q_B, M_B] = \frac{13}{27} \cdot \frac{24}{27} 0.16 + \frac{23}{27} \cdot \frac{30}{27} 0.25 = 0.3051 \ (\mathrm{kN})^2 \mathrm{m}$

Consider again the set of n random variables $\overline{X} = (X_1, \ldots, X_n)$ with expected values $E[X_i]$, $i = 1, 2, \ldots, n$ and variance and covariance collected in the covariance matrix $\overline{\overline{C}}_{\overline{X}}$ (defined by equation (2.84))

$$\overline{\overline{C}}_{\overline{X}} = \begin{bmatrix} \mathrm{Var}[X_1] & \mathrm{Cov}[X_1, X_2] & \cdots & \mathrm{Cov}[X_1, X_n] \\ \vdots & \vdots & & \vdots \\ \mathrm{Cov}[X_n, X_1] & \mathrm{Cov}[X_n, X_2] & \cdots & \mathrm{Var}[X_n] \end{bmatrix} \tag{6.9}$$

Clearly, no correlation between any pair of random variables will exist if this matrix $\overline{\overline{C}}_{\overline{X}}$ is a diagonal matrix. It will now be shown how a (new) set of random variables $\overline{Y} = (Y_1, \ldots, Y_n)$, where Y_i, $i = 1, 2, \ldots, n$ are linear functions of $X_1, \ldots, X_n$, can be constructed so that the corresponding covariance matrix $\overline{\overline{C}}_{\overline{Y}}$ is a diagonal matrix, that is

$$\overline{\overline{C}}_{\overline{Y}} = \begin{bmatrix} \mathrm{Var}[Y_1] & \cdots & 0 \\ \vdots & \ddots & \vdots \\ 0 & \cdots & \mathrm{Var}[Y_n] \end{bmatrix} \tag{6.10}$$

According to well-known theorems in linear algebra such a set of uncorrelated variables can be obtained by the transformation

$$\overline{Y} = \overline{\overline{A}}^T \overline{X} \tag{6.11}$$

where $\overline{\overline{A}}$ is an orthogonal matrix with column vectors equal to the orthonormal eigenvectors of $\overline{\overline{C}}_{\overline{X}}$. By this transformation

$$E[\overline{Y}] = \overline{\overline{A}}^T E[\overline{X}] \tag{6.12}$$

where $E[\overline{Y}] = (E[Y_1], \ldots, E[Y_n])$ and $E[\overline{X}] = (E[X_1], \ldots, E[X_n])$, and

$$\overline{\overline{C}}_{\overline{Y}} = \overline{\overline{A}}^T \overline{\overline{C}}_{\overline{X}} \overline{\overline{A}} \tag{6.13}$$

The diagonal elements of $\overline{\overline{C}}_{\overline{Y}}$, i.e. $\mathrm{Var}[Y_i]$, $i = 1, \ldots, n$ are equal to the eigenvalues of $\overline{\overline{C}}_{\overline{X}}$.

Example 6.3. Consider two correlated random variables X_1 and X_2 with the mean vector

$$E[\overline{X}] = (E[X_1], E[X_2]) = (2, 3)$$

and the covariance matrix

$$\bar{\bar{C}}_{\bar{X}} = \begin{bmatrix} Var[X_1] & Cov[X_1, X_2] \\ Cov[X_2, X_1] & Var[X_2] \end{bmatrix} = \begin{bmatrix} 3 & 1 \\ 1 & 3 \end{bmatrix}$$

The characteristic equation of $\bar{\bar{C}}_X$ is

$$(3 - \lambda)^2 - 1 = 0$$

with the roots $\lambda_1 = 2$ and $\lambda_2 = 4$. The corresponding orthonormal eigenvectors $\bar{v}_1$ and $\bar{v}_2$ are determined by the equations

$$\begin{bmatrix} 1 & 1 \\ 1 & 1 \end{bmatrix} \bar{v}_1 = \begin{bmatrix} 0 \\ 0 \end{bmatrix} \quad \text{and} \quad \begin{bmatrix} -1 & 1 \\ 1 & -1 \end{bmatrix} \bar{v}_2 = \begin{bmatrix} 0 \\ 0 \end{bmatrix}$$

resulting in

$$\bar{v}_1 = \frac{\sqrt{2}}{2}(1, -1) \quad \text{and} \quad \bar{v}_2 = \frac{\sqrt{2}}{2}(1, 1)$$

The transformation matrix $\bar{\bar{A}}$ is therefore given by

$$\bar{\bar{A}} = \frac{\sqrt{2}}{2}\begin{bmatrix} 1 & 1 \\ -1 & 1 \end{bmatrix}$$

The »new» uncorrelated random variables $\bar{Y} = (Y_1, Y_2)$ are therefore

$$\begin{bmatrix} Y_1 \\ Y_2 \end{bmatrix} = \frac{\sqrt{2}}{2}\begin{bmatrix} 1 & -1 \\ 1 & 1 \end{bmatrix}\begin{bmatrix} X_1 \\ X_2 \end{bmatrix}$$

or

$$Y_1 = \frac{\sqrt{2}}{2}(X_1 - X_2)$$
$$Y_2 = \frac{\sqrt{2}}{2}(X_1 + X_2)$$

The expected values are

$$E[Y_1] = \frac{\sqrt{2}}{2}(2 - 3) = -\frac{1}{2}\sqrt{2}$$
$$E[Y_2] = \frac{\sqrt{2}}{2}(2 + 3) = \frac{5}{2}\sqrt{2}$$

and the covariance matrix

$$\bar{\bar{C}}_{\bar{Y}} = \begin{bmatrix} \lambda_1 & 0 \\ 0 & \lambda_2 \end{bmatrix} = \begin{bmatrix} 2 & 0 \\ 0 & 4 \end{bmatrix} \qquad (6.14)$$

Note that $\bar{\bar{C}}_{\bar{Y}}$ can be calculated in the following way

$$\bar{\bar{C}}_{\bar{Y}} = \bar{\bar{A}}^T \bar{\bar{C}}_{\bar{X}} \bar{\bar{A}} = \frac{\sqrt{2}}{2}\begin{bmatrix} 1 & -1 \\ 1 & 1 \end{bmatrix}\begin{bmatrix} 3 & 1 \\ 1 & 3 \end{bmatrix}\frac{\sqrt{2}}{2}\begin{bmatrix} 1 & 1 \\ -1 & 1 \end{bmatrix}$$

but $\bar{\bar{C}}_{\bar{Y}}$ is of course easier determined by (6.14).

Exercise 6.2. Let the set of correlated random variables $\bar{X} = (X_1, X_2, X_3)$ have the means $E[\bar{X}] = [2, 4, 6]$ and the covariance matrix

$$\bar{\bar{C}}_{\bar{X}} = \begin{bmatrix} 4 & -1 & -1 \\ -1 & 7 & -2 \\ -1 & -2 & 7 \end{bmatrix}$$

Show that the random variables $\bar{Y} = (Y_1, Y_2, Y_3)$, where

$$\begin{bmatrix} Y_1 \\ Y_2 \\ Y_3 \end{bmatrix} = \frac{1}{\sqrt{6}}\begin{bmatrix} 2 & -\sqrt{2} & 0 \\ 1 & \sqrt{2} & -\sqrt{3} \\ 1 & \sqrt{2} & \sqrt{3} \end{bmatrix}\begin{bmatrix} X_1 \\ X_2 \\ X_3 \end{bmatrix}$$

are uncorrelated with the covariance matrix

$$\bar{\bar{C}}_{\bar{Y}} = \begin{bmatrix} 3 & 0 & 0 \\ 0 & 6 & 0 \\ 0 & 0 & 9 \end{bmatrix}$$

6.3 CORRELATED BASIC VARIABLES

In this section it is shown how the Hasofer and Lind reliability index β can be calculated for structures with correlated basic variables. It will be shown that it is often important to take account of this correlation. In section 5.4 the reliability index β was defined in the following way. First a set of *uncorrelated basic variables* $\bar{X} = (X_1, \ldots, X_n)$ is chosen, and the failure surface is expressed in the corresponding x-coordinate system. Then a new set of variables $\bar{Z} = (Z_1, \ldots, Z_n)$ is obtained by normalizing the original set of basic variables $\bar{X}$ in the following way

$$Z_i = \frac{X_i - \mu_{X_i}}{\sigma_{X_i}} \quad , \; i = 1, 2, \ldots, n \tag{6.15}$$

where μ_{X_i} and σ_{X_i} are the mean and the standard deviation of X_i. By the linear mapping (6.15) the failure surface in the x-coordinate system is mapped into a failure surface in the z-coordinate system. The reliability index β is then defined as the shortest distance from the origin to the failure surface in the normalized z-coordinate system.

When the *basic variables* $\overline{X} = (X_1, \ldots, X_n)$ *are correlated* the first step is to obtain a set of *uncorrelated variables* $\overline{Y} = (Y_1, \ldots, Y_n)$ as shown in the last section. The next step is then to normalize the uncorrelated variables and thereby obtain a set of *normalized and uncorrelated variables* $\overline{Z} = (Z_1, \ldots, Z_n)$. Finally, the reliability index can be defined in the z-coordinate system as shown above. This procedure is illustrated in example 6.4.

Example 6.4. Consider the simply supported beam shown on figure 6.3. The beam is loaded with two concentrated loads X_1 and X_2 with the means $E[X_1] = E[X_2] = \mu$ and the covariance matrix

$$\overline{\overline{C}}_{\overline{X}} = \begin{bmatrix} \mathrm{Var}[X_1] & \mathrm{Cov}[X_1, X_2] \\ \mathrm{Cov}[X_2, X_1] & \mathrm{Var}[X_2] \end{bmatrix} = \sigma^2 \begin{bmatrix} 1 & \rho \\ \rho & 1 \end{bmatrix}$$

where $-1 < \rho < 1$.

Uncorrelated variables Y_1 and Y_2 are according to equation (6.11) determined by

$$\overline{Y} = \overline{\overline{A}}^T \overline{X}$$

where the transformation matrix is determined by the eigenvectors for $\overline{\overline{C}}_{\overline{X}}$. The eigenvalues λ_1, λ_2 and eigenvectors $\overline{v}_1, \overline{v}_2$ for $\overline{\overline{C}}_{\overline{X}}$ are easily determined in this simple case

$$\lambda_1 = \sigma^2(1+\rho) \qquad \overline{v}_1 = \frac{\sqrt{2}}{2}(1, 1)$$

$$\lambda_2 = \sigma^2(1-\rho) \qquad \overline{v}_2 = \frac{\sqrt{2}}{2}(1, -1)$$

The transformation $\overline{\overline{A}}$ is therefore given by

$$\overline{\overline{A}}^T = \frac{\sqrt{2}}{2} \begin{bmatrix} 1 & 1 \\ 1 & -1 \end{bmatrix}$$

and, according to the equations (6.11) - (6.13),

$$\overline{Y} = \overline{\overline{A}}^T \overline{X} \Rightarrow \begin{cases} Y_1 = \frac{\sqrt{2}}{2}(X_1 + X_2) \\ Y_2 = \frac{\sqrt{2}}{2}(X_1 - X_2) \end{cases}$$

$$E[\overline{Y}] = \overline{\overline{A}}^T E[\overline{X}] \Rightarrow \begin{cases} E[Y_1] = \sqrt{2}\,\mu \\ E[Y_2] = 0 \end{cases}$$

$$\overline{\overline{C}}_{\overline{Y}} = \overline{\overline{A}}^T \overline{\overline{C}}_{\overline{X}} \overline{\overline{A}} \Rightarrow \overline{\overline{C}}_{\overline{Y}} = \sigma^2 \begin{bmatrix} 1+\rho & 0 \\ 0 & 1-\rho \end{bmatrix}$$

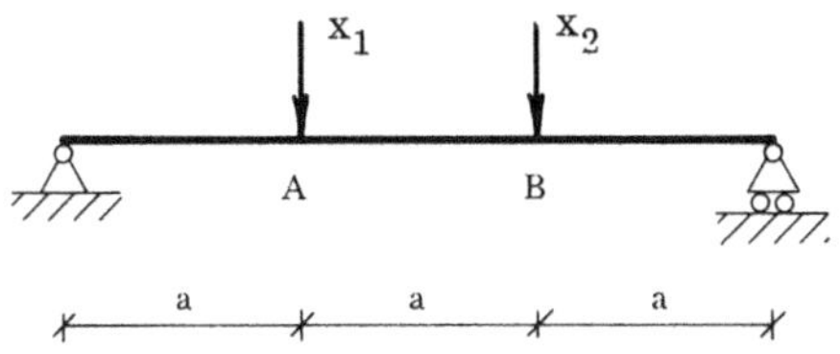

Figure 6.3

Uncorrelated and normalized variables Z_1 and Z_2 are finally determined by

$$Z_1 = \frac{Y_1 - E[Y_1]}{\sigma_{Y_1}} = \frac{\sqrt{2}}{2}(X_1 + X_2 - 2\mu)/\sigma\sqrt{1+\rho} \tag{6.16}$$

$$Z_2 = \frac{Y_2 - E[Y_2]}{\sigma_{Y_2}} = \frac{\sqrt{2}}{2}(X_1 - X_2)/\sigma\sqrt{1-\rho} \tag{6.17}$$

From the matrix equations (6.11) - (6.13) it is seen that

$$\overline{Z} = \overline{\overline{C}}_{\overline{Y}}^{\,-\frac{1}{2}}(\overline{Y} - E[\overline{Y}])$$

or

$$\overline{Z} = (\overline{\overline{A}}^T \overline{\overline{C}}_{\overline{X}} \overline{\overline{A}})^{-\frac{1}{2}} \; \overline{\overline{A}}^T (\overline{X} - E[\overline{X}]) \tag{6.18}$$

Note that in the z-coordinate system we have

$$\overline{z}^T \overline{z} = (\overline{x} - E[\overline{X}])^T \overline{\overline{A}} (\overline{\overline{A}}^T \overline{\overline{C}}_{\overline{X}} \overline{\overline{A}})^{-\frac{1}{2}} (\overline{\overline{A}}^T \overline{\overline{C}}_{\overline{X}} \overline{\overline{A}})^{-\frac{1}{2}} \overline{\overline{A}}^T (\overline{x} - E[\overline{X}]) =$$

$$(\overline{x} - E[\overline{X}])^T \overline{\overline{A}} (\overline{\overline{A}}^T \overline{\overline{C}}_{\overline{X}} \overline{\overline{A}})^{-1} \; \overline{\overline{A}}^T (\overline{x} - E[\overline{X}]) = (\overline{x} - E[\overline{X}])^T \overline{\overline{C}}_{\overline{X}}^{-1} (\overline{x} - E[\overline{X}]) \tag{6.19}$$

The definition of the reliability index β by Hasofer and Lind can now be extended to

$$\beta = \min_{\overline{z} \in \omega_z} (\overline{z}^T \overline{z})^{\frac{1}{2}} = \min_{\overline{x} \in \omega_x} ((\overline{x} - E[\overline{X}])^T \overline{\overline{C}}_{\overline{X}}^{-1} (\overline{x} - E[\overline{X}]))^{\frac{1}{2}} \tag{6.20}$$

where ω_z is the failure surface in the z-coordinate system and ω_x the failure surface in the x-coordinate system.

The formal definition (6.20) is convenient if a computer is available, because calculation of β is by (6.20) formulated as a constrained optimization problem. The constrained Fletcher-Powell technique can be used for this purpose. For problems with a small number of basic variables the same iterative method as used in the examples in chapter 5 can be used. A small desk computer or pocket computer is suitable for this purpose.

It is easy to see that the reliability index β defined by equation (6.20) can be evaluated on the basis of a safety margin, when the failure function is linear in the basic variable. Let the failure function $\mathbf{f}: R^n \to R$ be given by

$$f(\bar{x}) = a_0 + a_1 x_1 + \ldots + a_n x_n \tag{6.21}$$

and the corresponding safety margin

$$M_X = a_0 + a_1 X_1 + \ldots + a_n X_n$$

The reliability index β is then simply equal to

$$\beta = \frac{\mu_{M_X}}{\sigma_{M_X}} \tag{6.22}$$

i.e. the same value as determined on the basis of the reliability index β as defined by equation (5.9). An illustration of this is given in example 6.5 for the 2-dimensional case.

Example 6.5. Let X_1 and X_2 be random variables with means $E[X_1] = E[X_2] = \mu$ and the covariance matrix

$$\bar{\bar{C}}_{\bar{X}} = \sigma^2 \begin{bmatrix} 1 & \rho \\ \rho & 1 \end{bmatrix}$$

exactly as in example 6.4. Further, let the safety margin M_X be defined by

$$M_X = a_0 + a_1 X_1 + a_2 X_2 \tag{6.23}$$

The purpose of this example is to show that (6.20) and (6.22) yield the same value for the reliability index β.

Uncorrelated normalized variables Z_1 and Z_2 can be calculated from equations (6.16) and (6.17). The inverse equations are

$$X_1 = \mu + \frac{\sqrt{2}}{2} (\sigma\sqrt{1+\rho}\, Z_1 + \sigma\sqrt{1-\rho}\, Z_2)$$

$$X_2 = \mu + \frac{\sqrt{2}}{2} (\sigma\sqrt{1+\rho}\, Z_1 - \sigma\sqrt{1-\rho}\, Z_2)$$

Therefore, the safety margin M_Z can be written

$$M_Z = a_0 + \mu(a_1 + a_2) + \frac{\sqrt{2}}{2}\sigma\sqrt{1+\rho}\,(a_1 + a_2) Z_1 + \frac{\sqrt{2}}{2}\sigma\sqrt{1-\rho}\,(a_1 - a_2) Z_2 \tag{6.24}$$

or

$$M_Z = b_0 + b_1 Z_1 + b_2 Z_2 \tag{6.25}$$

The reliability index of Hasofer and Lind defined by (6.20) is equal to the distance from the origin to the straight line (6.25), i.e.

$$\beta = \frac{b_0}{\sqrt{b_1^2 + b_2^2}} \tag{6.26}$$

where $b_0 = a_0 + \mu(a_1 + a_2)$ and $b_1^2 + b_2^2 = \sigma^2(a_1^2 + a_2^2 + 2\rho a_1 a_2)$, so that

$$\beta = \frac{a_0 + \mu(a_1 + a_2)}{\sigma\sqrt{a_1^2 + a_2^2 + 2\rho a_1 a_2}} \tag{6.27}$$

From (6.23)

$$\mu_{M_X} = a_0 + (a_1 + a_2)\mu$$

and

$$\sigma_{M_X}^2 = a_1^2\sigma^2 + a_2^2\sigma^2 + 2a_1 a_2 \rho\sigma^2 = \sigma^2(a_1^2 + a_2^2 + 2a_1 a_2 \rho)$$

Therefore, (6.22) gives the same result as (6.27).

Example 6.6. To illustrate the importance of taking the correlation into account consider example 6.5 with $a_0 = 0$ and $a_1 = a_2 = 1$, i.e. with the safety margin

$$M_X = X_1 + X_2$$

Then, from (6.27) it follows that

$$\beta = (1 + \rho)^{-\frac{1}{2}}\beta^* \tag{6.28}$$

where β^* corresponds to no correlation between X_1 and X_2 ($\rho = 0$). This relationship between ρ and β/β^* is shown in figure 6.4.

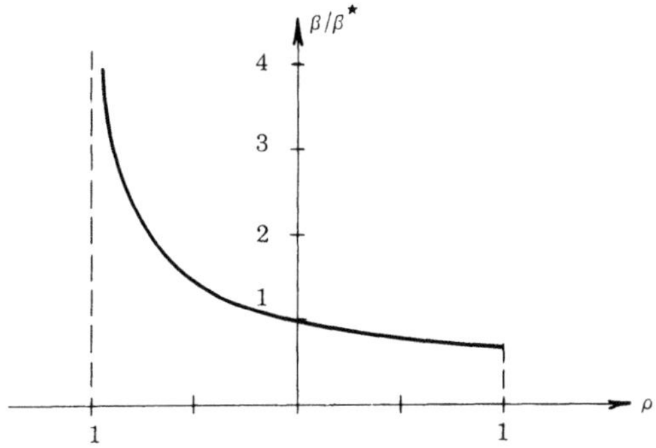

Figure 6.4

Exercise 6.3. Consider the beam shown in figure 6.5 loaded with two concentrated loads p_1 and p_2 at the same point on the beam. The maximum deflection is

$$u_{max} = \frac{1}{9\sqrt{3}} \frac{pc\ell^2}{ei}$$

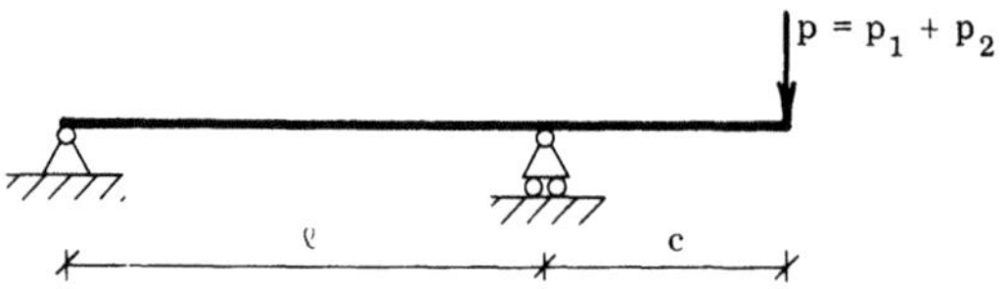

Figure 6.5

where e is the modulus of elasticity, i the relevant moment of inertia and $p = p_1 + p_2$. Further, let p, c, ℓ, e and i be realizations of random variables $P = P_1 + P_2$, C, L, E and I with

$$E[P_1] = E[P_2] = 8 \text{ kN} \qquad \sigma_{P_1} = \sigma_{P_2} = 0.32 \text{ kN}$$

$$E[C] = 2 \text{ m} \qquad \sigma_C \sim 0 \text{ m}$$

$$E[L] = 4 \text{ m} \qquad \sigma_L \sim 0 \text{ m}$$

$$E[E] = 4 \cdot 10^{-5} \text{ m}^{-4} \qquad \sigma_E = 0.3 \cdot 10^{-5} \text{ m}^4$$

$$E[I] = 4 \cdot 10^4 \text{ MN/m}^2 \qquad \sigma_I = 10^4 \text{ MN/m}^2$$

All random variables except P_1 and P_2 are assumed uncorrelated. The correlation between P_1 and P_2 is given by

$$\text{Cov}[P_1, P_2] = 0.32 \text{ (kN)}^2$$

Calculate the reliability index β when the following failure criterion is used

$$u_{max} \geqslant \frac{1}{100} \ell$$

In examples 6.5 - 6.6 and exercise 6.3 the correlation between random variables could be treated in a simple way because these random variables only appeared in a linear connection. The important equation (6.5) could therefore be used in these examples. It was not necessary to construct new uncorrelated variables in the way presented on page 99. This is not the case in the next example, where the failure function is non-linear in the two basic random variables which are also correlated.

Example 6.7. Consider again the beam shown in figure 6.5, but now only c and ℓ are considered realizations of random variables. All the other variables are assumed to be deterministic. The maximum deflection can therefore be written

$$u_{max} = kc\ell^2$$

where k is a constant. Further assume that c and ℓ are realizations of random variables with $E[L] = 4$ m, $E[C] = 2$ m, $\sigma_L = \sigma_C = 0.25$ m and $Cov[L, C] = 1/32$ m^2. The covariance matrix is then

$$\bar{\bar{C}} = \frac{1}{16}\begin{bmatrix} 1 & 0.5 \\ 0.5 & 1 \end{bmatrix}$$

so that the results from example 6.4 can be used directly. Uncorrelated random variables Y_1 and Y_2 are given by

$$Y_1 = \frac{\sqrt{2}}{2}(L + C)$$
$$Y_2 = \frac{\sqrt{2}}{2}(L - C) \tag{6.29}$$

with $E[Y_1] = 3\sqrt{2}$ m, $E[Y_2] = \sqrt{2}$ m and the covariance matrix

$$\bar{\bar{C}}_{\bar{Y}} = \frac{1}{16}\begin{bmatrix} 1.5 & 0 \\ 0 & 0.5 \end{bmatrix} \tag{6.30}$$

From (6.29) the random variables L and C can be expressed by Y_1 and Y_2

$$L = \frac{\sqrt{2}}{2}(Y_1 + Y_2)$$
$$C = \frac{\sqrt{2}}{2}(Y_1 - Y_2) \tag{6.31}$$

Let $k = 6.415 \cdot 10^{-4}$ and let the failure criterion be

$$u_{max} \geqslant \frac{1}{100}\ell$$

or

$$1 - 6.415 \cdot 10^{-2} c\ell \leqslant 0 \tag{6.32}$$

From (6.31) it follows that

$$LC = \frac{1}{2}(Y_1^2 - Y_2^2)$$

so that the failure criterion (6.32) can be written

$$1 - 3.208 \cdot 10^{-2}(y_1^2 - y_2^2) \leqslant 0 \tag{6.33}$$

where y_1 and y_2 are realizations of uncorrelated random variables Y_1 and Y_2. The reliability index β can now be calculated by the iterative method shown in Chapter 5.

: **Exercise 6.4.** Determine the reliability index β for the safety problem formulated in exam-
: ple 6.7.

6.4 NON-NORMAL BASIC VARIABLES

Until now only second order information has been taken into account when evaluating the reliability of a structure. Let the failure surface be linear and let the basic variables be normally distributed. Then the following relation between the probability of failure P_f and the reliability index β exists (see (5.33))

$$P_f = \Phi(-\beta) \Leftrightarrow \beta = -\Phi^{-1}(P_f) \tag{6.34}$$

where Φ is the standardized normal distribution function.

It is important to remember that no information regarding the probability of failure can be obtained when the distributions of the basic variables are unknown. It is reasonable in some situations to expect that a given random variable can be considered normally distributed as a good approximation, but often such an assumption is quite unreasonable. Consider the yield stress of a steel bar. For such a random variable negative values cannot occur. It is therefore more realistic to assume log-normal distribution in this case. To overcome the problem that the definition space for the set of basic variables $X_1, \dots, X_n$ is not the whole R^n space, one can apply one-to-one transformations to the relevant variables. For the yield stress mentioned above this transformation should map R_+ on R and be continuous.

In this chapter the transformation given by Rackwitz and Fiessler [6.4] will be used. This transformation is chosen in such a way that the values of the original density functions f_{X_i} and the original distribution function F_{X_i} for the random variables X_i are equal to the corresponding values of the density function and the distribution function for a normally distributed variable at the design point A (see figure 5.5), i.e.

$$F_{X_i}(x_i^*) = \Phi\left(\frac{x_i^* - \mu'_{X_i}}{\sigma'_{X_i}}\right) \tag{6.35}$$

$$f_{X_i}(x_i^*) = \frac{1}{\sigma'_{X_i}} \varphi\left(\frac{x_i^* - \mu'_{X_i}}{\sigma'_{X_i}}\right) \tag{6.36}$$

where the design point $A = (x_1^*, \dots, x_i^*, \dots, x_n^*)$ and where μ'_{X_i} and σ'_{X_i} are the (unknown) mean and standard deviation of the approximate normal distribution. Solving (6.35) and (6.36) with regard to μ'_{X_i} and σ'_{X_i} we have

$$\sigma'_{X_i} = \frac{\varphi(\Phi^{-1}(F_{X_i}(x_i^*)))}{f_{X_i}(x_i^*)} \tag{6.37}$$

$$\mu'_{X_i} = x_i^* - \Phi^{-1}(F_{X_i}(x_i^*))\sigma'_{X_i} \tag{6.38}$$

Clearly the iterative method presented in chapter 5 for calculating the reliability index β for a given structure must be modified a little when the transformation shown above is used. On each step of the iteration new values for σ'_{X_i} and μ'_{X_i} must be calculated for those variables where such a transformation has been used. This is illustrated in example 6.8.

Example 6.8. Consider again the beam analysed in example 5.5. In example 5.5 the reliability index β was calculated solely on the basis of second order moments for the relevant basic variables, namely the load P, the modulus of elasticity E and the moment of inertia I. It will now be assumed that the load P is Gumbel distributed with the distribution function (see (3.10))

$$F_P(p) = \exp(-\exp(-\alpha(p-u)))$$

and the density function

$$f_P(p) = \exp[-\exp(-\alpha(p-u)) - \alpha(p-u)]\alpha$$

The two parameters α and u can be calculated from the following expressions for the mean μ_P and the standard deviation σ_P (see (3.11) and (3.12))

$$\mu_P = u + 0.57722/\alpha$$

$$\sigma_P = \frac{\pi}{\sqrt{6}} \cdot \frac{1}{\alpha}$$

when μ_P and σ_P are known. With the same mean and standard deviation for P, E and I as in example 5.5 one gets $\alpha = 1.2825\ (\text{kN})^{-1}$ and $u = 3.5499$ kN. The Gumbel distribution is now transformed into a normal distribution with the mean μ'_P and the standard deviation σ'_P given by the equations (6.37) and (6.38).

In the normalized coordinate system the failure surface is then (compare with example 5.5)

$$(10^{-4} + 0.2 \cdot 10^{-4} x_1)(2 \cdot 10^7 + 0.5 \cdot 10^7 x_2) - 78.12(\mu'_P + \sigma'_P x_3) = 0 \qquad (6.39)$$

The reliability index β can now be calculated by an iterative technique analogous with the method used in example 5.5. The only difference is that the new values of μ'_P and σ'_P must be calculated after each step in the iteration in the following way

$$\mu'_P = x_3^* - \Phi^{-1}(F_P(x_3^*))\sigma'_P \qquad (6.40)$$

$$\sigma'_P = \varphi(\Phi^{-1}(F_P(x_3^*)))/f_P(x_3^*) \qquad (6.41)$$

where $x_3^* = F_P^{-1}(\Phi(\beta\alpha_3))$.

	Start	Iteration No.					
		1	2	3	4	5	6
β	3	3.53	3.50	3.40	3.34	3.33	3.32
α_1	−0.58	−0.52	−0.35	−0.24	−0.20	−0.18	−0.18
α_2	−0.58	−0.78	−0.88	−0.94	−0.95	−0.96	−0.97
α_3	0.58	0.35	0.31	0.26	0.22	0.20	0.20
σ'_P	1.708	1.425	1.337	1.237	1.176	1.150	1.140
μ'_P	3.054	3.479	3.580	3.678	3.728	3.746	3.752

Table 6.1

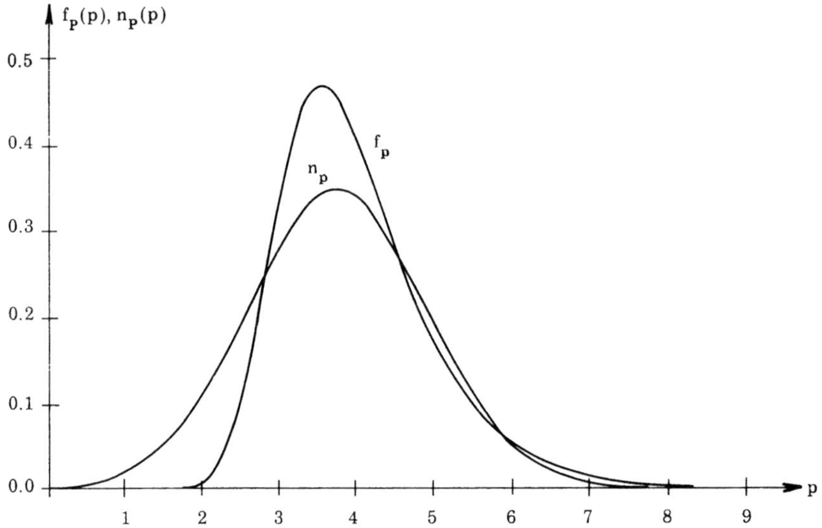

Figure 6.6

The reliability index is now $\beta = 3.32$ which is a small increase from example 5.5, where $\beta = 3.29$.
The Gumbel density function f_P and the corresponding normal density function n_P are shown in figure 6.6.
According to the equations (6.35) and (6.36) n_P and the corresponding distribution function N_P are connected with f_P and F_P in the following way

$$n_P(x_3^\star) = f_P(x_3^\star) \tag{6.42}$$

$$N_P(x_3^\star) = F_P(x_3^\star) \tag{6.43}$$

where $x_3^\star$ is the third coordinate of the design point

$$(x_1^\star, x_2^\star, x_3^\star) = \beta(\alpha_1, \alpha_2, \alpha_3) = (-0.59, -3.22, 0.66)$$

Exercise 6.5. Consider again the elastic beam treated in example 5.6, but assume now that the load P is Gumbel distributed with the mean $\mu_P = 4$ kN and the standard deviation $\sigma_P = 1$ kN. The other data used in example 5.6 are unchanged.

Map this Gumbel distribution on a normal distribution as shown above and determine the expected value μ_I so that $\beta = 3$ (solution $\mu_I = 161 \cdot 10^{-7}\ \text{m}^4$).

BIBLIOGRAPHY

[6.1] Ditlevsen, O.: *Fundamentals of Second Moment Structural Reliability Theory*. Int. Research Seminar on Safety of Structures, Trondheim, Norway, 1977.

[6.2] Ditlevsen, O.: *Uncertainty Modeling*. McGraw-Hill, 1981.

[6.3] Hasofer, A. M. and N. Lind: *An Exact and Invariant First Order Reliability Format.* Proc. ASCE, J. Eng. Mech. Div., 1974, pp. 111 - 121.

[6.4] Rackwitz, R. and B. Fiessler: *An Algorithm for Calculation of Structural Reliability under Combined Loading.* Berichte zur Sicherheitstheorie der Bauwerke, Lab. f. Konstr. Ingb., München, 1977.

[6.5] *Rationalisation of Safety and Serviceability Factors in Structural Codes.* Report 63, Construction Industry Research and Information Association, London, England, 1977.

[6.6] The Nordic Committee on Building Regulations (NKB): *The Loading and Safety Group, Recommendations for Loading and Safety Regulations for Structural Design.* NKB-Report No. 36, November 1978.

[6.7] Thoft-Christensen, P.: *Introduction to Reliability of Offshore Structures.* Lectures on Offshore Engineering (eds. W. J. Graff and P. Thoft-Christensen), Aalborg University Centre, Aalborg, Denmark, 1978, pp. 53 - 72.

[6.8] Thoft-Christensen, P.: *Some Experience from Application of Optimization Technique in Structural Reliability.* Bygningsstatiske Meddelelser, Vol. 48, 1977, pp. 31 - 44.

Chapter 7

RELIABILITY OF STRUCTURAL SYSTEMS

7.1 INTRODUCTION

In the preceding chapters the concept of reliability has mainly been concerned with single structural elements such as a beam or a column. In the fundamental case the loading is described by a single random variable S and the strength by a single random variable R (see chapter 4). The probability of failure P_f is then defined as

$$P_f = P(S \geqslant R) \tag{7.1}$$

assuming that the failure condition is $R - S \leqslant 0$. P_f can be calculated from

$$P_f = \int_{\omega_f} f_{R,S}(r, s)\, drds \tag{7.2}$$

where $f_{R,S}$ is the joint probability density function and ω_f the failure region $\{(r, s) | r - s \leqslant 0\}$. When $f_{R,S}$ is known, the probability of failure P_f can be calculated relatively easily from (7.2) by a suitable numerical technique or by simulation.

When the loading of a single structural element is determined by a number of random variables $\bar{R} = (R_1, \ldots, R_n)$ and likewise the strength by a number of random variables $\bar{S} = (S_1, \ldots, S_m)$ equation (7.2) can be generalised and the probability of failure calculated if the corresponding joint probability density function is known.

In the situations described above only one single structural member with a single failure mode is treated. The reliability of a real structure is usually much more difficult to evaluate since more than one element (member) can fail and because there is possibility of more than one failure mode for the system. To handle problems of this kind it is sometimes useful to consider from a systems point of view. In this the real structure is modelled by an equivalent system in such a way that all relevant failure modes can be treated.

In this chapter the fundamental systems - series systems and parallel systems - will be defined and it will be shown how such systems can be analysed from a probabilistic point of view, both when no correlation between the elements exists and when all elements are equally correlated. In Chapter 8 some important reliability bounds for structural systems will be shown, and for systems with unequally correlated elements some approximate methods of estimating the failure probability will be presented.

7.2 PERFECTLY BRITTLE AND PERFECTLY DUCTILE ELEMENTS

It is of great importance for a structural system whether its elements can be considered *perfectly brittle* or *perfectly ductile.* A structural element is called *perfectly brittle,* if it becomes ineffective after failure, i.e. if it loses its load-bearing capacity completely by failure.

The words »perfectly brittle» should be understood in a broad sense. If a tensile bar made of a brittle material fails due to a tensile force then such an element can reasonably be considered perfectly brittle, because its loading capacity is completely exhausted. But the characterization »perfectly brittle» can also be used in situations, where no real fracture occurs, e.g. when the element is deflected so much that it is ineffective in relation to a given loading. A typical load-deflection curve for a brittle element is shown in figure 7.1.

If an element maintains its load level after failure it is called *perfectly ductile.* A typical example of perfectly ductile behaviour is shown in figure 7.2, where it is assumed that the load p can be maintained during an increasing displacement. To distinguish these two types of element behaviour the symbols shown in figure 7.3 can be used.

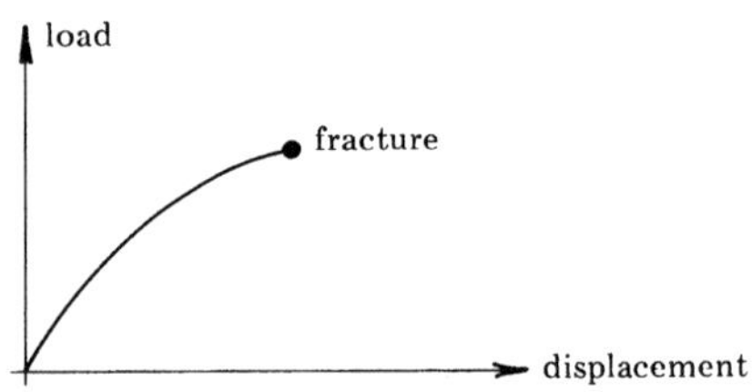

Figure 7.1

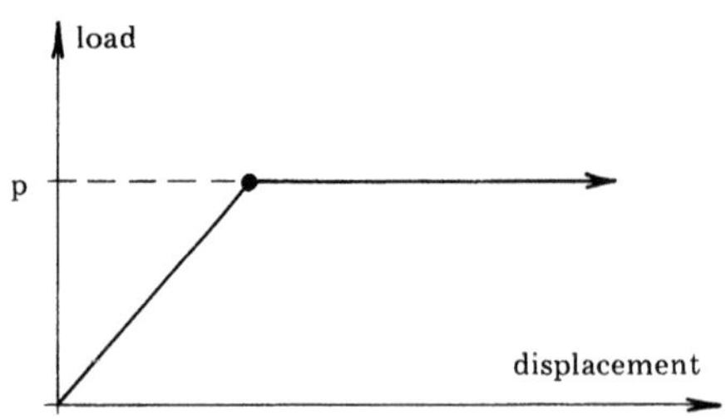

Figure 7.2

Figure 7.3

7.3 FUNDAMENTAL SYSTEMS

As mentioned in the introduction there are two fundamental types of systems, namely series systems and parallel systems. A system of single elements is a *series system* if it is in a state of failure whenever any of its elements fails. Such a system is also called a *weakest-link system*. A typical example of a series system is a statically determinate structure as shown in figure 7.4. Obviously, failure in any member of such a structure will result in failure of the total system. A series system with n elements is generally symbolised as shown in figure 7.5.

All elements in figure 7.5 are brittle elements but for a series system the distinction between brittle and ductile elements is irrelevant because the total system fails as soon as one element fails whether it is brittle or ductile. It is important to note that the idealisation of a structure by a series system as in figure 7.5 is only related to the failure interaction. Therefore, one must not interpret figure 7.5 as one in which all elements have the same load, although a load S is shown applied to each end of the series system. Usually, an external load on a series system

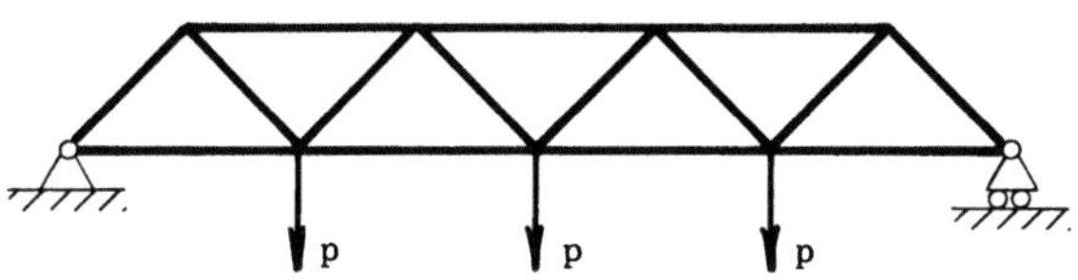

Figure 7.4

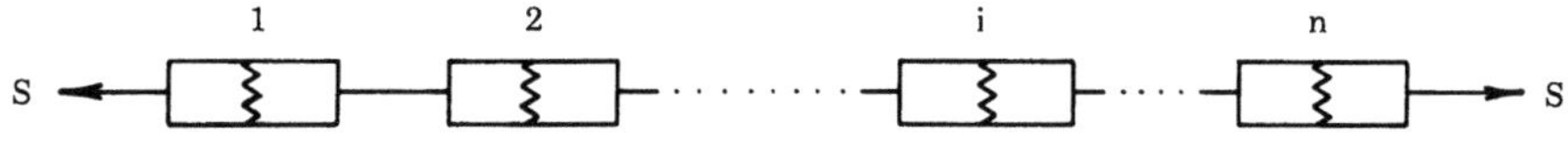

Figure 7.5

will result in different loads (stresses) in the single elements. Figure 7.5 is useful in calculating the distribution function F_R for the strength R of the series system. Let F_{R_i} be the distribution function for the strength R_i of element i, then

$$F_R(r) = P(R \leqslant r) = 1 - P(R > r) = 1 - P(R_1 > r_1 \cap R_2 > r_2 \cap \ldots \cap R_n > r_n)$$

$$= 1 - (1 - F_{R_1}(r_1))(1 - F_{R_2}(r_2)) \ldots (1 - F_{R_n}(r_n)) = 1 - \prod_{i=1}^{n} (1 - F_{R_i}(r_i)) \qquad (7.3)$$

where it is assumed that the strengths of the elements are independent.

Example 7.1. Consider a simple structure consisting of two tensile bars as shown in figure 7.6. Let the strength R_e of each tensile bar be a random variable with the density function f_{R_e} shown in the same figure. Let the structure be loaded by a single tensile force S = 1.1 kN. It is reasonable to model this structure as a series system with two elements. The distribution function F_R for the strength R of the system can then be derived by equation (7.3) if the strength of two elements can be assumed independent. One obtains

$$F_R(r) = P(R \leqslant r) = 1 - (1 - F_{R_e}(r))^2$$

where

$$F_{R_e}(r) = \begin{cases} 0 & \text{for} \quad r < 1 \\ \frac{1}{2} r - \frac{1}{2} & \text{for} \quad 1 \leqslant r < 3 \\ 1 & \text{for} \quad r \geqslant 3 \end{cases}$$

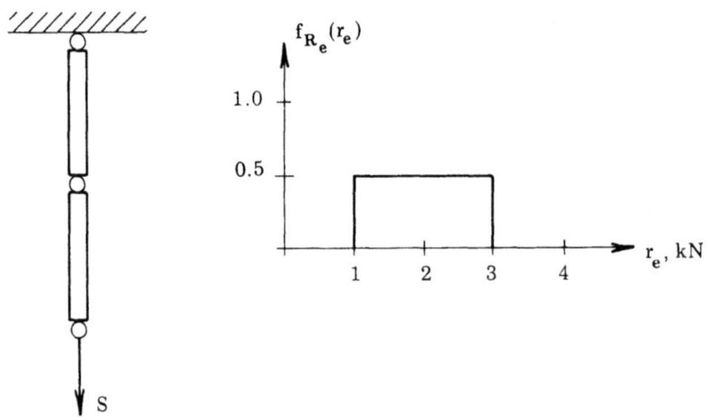

Figure 7.6

: By substituting F_{R_e} in the expression for F_R one obtains

$$F_R(r) = \begin{cases} 0 & \text{for } r < 1 \\ -\frac{1}{4}r^2 + \frac{3}{2}r - \frac{5}{4} & \text{for } 1 \leqslant r < 3 \\ 1 & \text{for } r \geqslant 3 \end{cases}$$

: It is then easy to calculate the mean μ_R and the variance σ_R^2 for the strength R of the system. One gets

$$\mu_R = \frac{5}{3} \quad \text{and} \quad \sigma_R^2 = \frac{2}{9}$$

: The reliability index β_S for the system is therefore

$$\beta_S = \frac{\mu_R - \mu_S}{\sqrt{\sigma_R^2 + \sigma_S^2}} = \frac{5/3 - 1.1}{\sqrt{2/9}} = 1.24$$

: while the reliability index β_e for an element is

$$\beta_e = \frac{\mu_{R_e} - \mu_S}{\sqrt{\sigma_{R_e}^2 + \sigma_S^2}} = \frac{2 - 1.1}{\sqrt{1/3}} = 1.56$$

: Note that $\beta_S < \beta_e$ as expected. Also note that better values can be calculated on the basis of (7.4) and (6.34).

When the distribution function F_R for the strength R of the series system is determined, the probability of failure P_f can be calculated as for a single element by

$$P_f = \int_{-\infty}^{\infty} F_R(r) f_S(r) dr = 1 - \int_{-\infty}^{\infty} \prod_{i=1}^{n} (1 - F_{R_i}(r_i)) f_S(r) dr \tag{7.4}$$

where f_S is the density function for the load S on the series system.

: **Example 7.2.** Consider again the structure shown in figure 7.6 loaded with a deterministic force S = 1.1 kN. In this case the probability of failure is simply

$$P_f = F_R(1.1) = -\frac{1}{4} \cdot 1.21 + 1.5 \cdot 1.1 - 1.25 = 0.0975$$

: while the probability of failure for a single element is

$$P_{f_e} = F_{R_e}(1.1) = \frac{1}{2} \cdot 1.1 - 0.5 = 0.05$$

: **Exercise 7.1.** To illustrate that the probability of failure P_f of a series system increases
: with the number of elements n, consider a system with n elements with the same distribu-
: tion function F_{R_e} as used in example 7.1 and the same force S = 1.1 kN. Calculate P_f for
: n = 2, 5 and 10.

Now turn to *parallel systems.* Failure in a single element in a structural system will not always result in failure of the total system, because the remaining elements may be able to sustain the external loads by redistribution of the loads. This situation is a characteristic of statically indeterminate structures. Failure of such structures will always require that more than one element fails. From a strength (failure) point of view such a set of elements is called a parallel system and the associated failure state is called a failure mode.

A real statically indeterminate structure will usually have a great number of failure modes, where each failure mode is modelled by a parallel system, and these parallel systems are then again combined as a series system. Therefore, such a structural system will fail when the weakest mode (parallel system) fails. In other words, a parallel system will only fail when all elements in that system fail. Therefore, the behaviour of such a system depends to a high degree on whether the elements are perfectly ductile or perfectly brittle.

A *parallel system with n perfectly ductile elements* is shown in figure 7.7. Because of the assumption of ductile elements, the strength R of this system is simply determined by

$$R = \sum_{i=1}^{n} R_i \tag{7.5}$$

where the strength of element i is given by the random variable R_i. Note that when the random variables R_i, i = 1, 2, . . . , n are independent and are normally distributed $N(\mu_i, \sigma_i)$, then R is also normally distributed $N(\mu, \sigma)$, with

$$E[R] = \mu = \sum_{i=1}^{n} \mu_i \tag{7.6}$$

$$Var[R] = \sigma_R^2 = \sum_{i=1}^{n} \sigma_i^2 \tag{7.7}$$

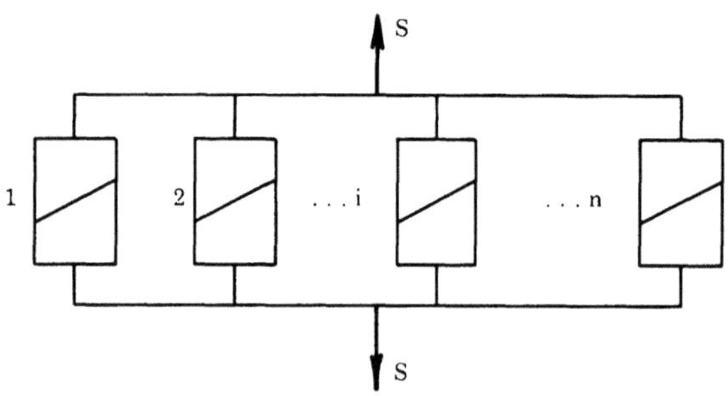

Figure 7.7

According to the Central Limit Theorem it is reasonable to assume R to be normally distributed if the number of elements is not too small, even in cases where the distributions of R_i, $i = 1, 2, \ldots, n$, are non-normal.

Example 7.3. Consider the system shown in figure 7.8 with 3 elements with the strengths R_1, R_2 and R_3. Let R_1 and R_2 be normally distributed with $E[R_1] = E[R_2] = 5$ kN and $\sigma_{R_1} = \sigma_{R_2} = 1$ kN. Further, let R_3 have a uniform distribution over the interval [8 kN ; 12 kN]. The random variables are assumed statistically independent.

Consider first the subsystem consisting of the two elements 1 and 2. According to (7.5) the strength R_{12} of this subsystem is normally distributed with $E[R_{12}] = 10$ kN and $\sigma R_{12} = \sqrt{2}$ kN. The total system can now be considered a series sytems with two elements, namely an element with the strength R_{12} normally distributed and an element with the strength R_3 uniformly distributed. The distribution function F_{R_3} for R_3 is given by

$$F_{R_3}(r) = \begin{cases} 0 & \text{for } r < 8 \\ \frac{1}{4}r - 2 & \text{for } 8 \leqslant r < 12 \\ 1 & \text{for } 12 \leqslant r \end{cases}$$

The distribution function F_R of the total system can now be calculated from (7.3)

$$F_R(r) = 1 - (1 - \Phi(\frac{r-10}{\sqrt{2}}))(1 - F_{R_3}(r))$$
$$= \Phi(\frac{r-10}{\sqrt{2}}) + F_{R_3}(r) - \Phi(\frac{r-10}{\sqrt{2}}) F_{R_3}(r)$$

where Φ is the standard normal distribution function.

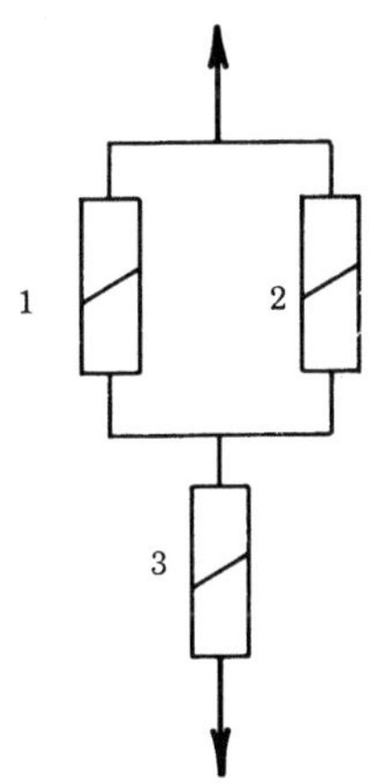

Figure 7.8

Exercise 7.2. Consider the structural system shown in figure 7.9 consisting of 3 tensile bars. Assume that this structural system can be modelled by the system shown in figure 7.8 and let its strength be as calculated in example 7.3. Determine the probability of failure P_f for this system when it is loaded with a single force S = 7.5 kN. What is P_f when S = 10 kN?

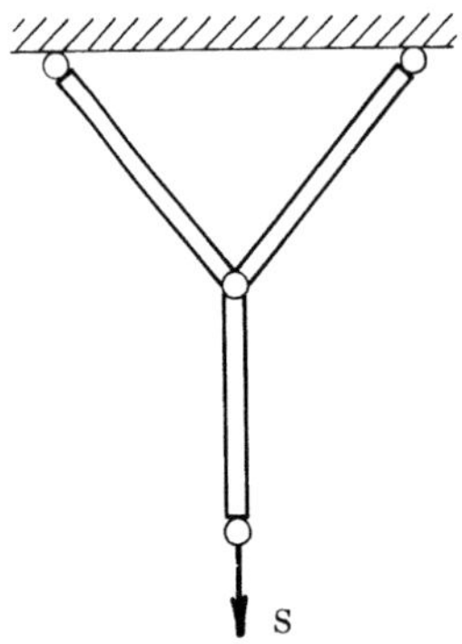

Figure 7.9

Parallel ductile systems are relatively easy to deal with because of the simple relation (7.5) between the strength R of the system and the strengths R_i, $i = 1, 2, \ldots, n$ of the individual elements. A parallel system with perfectly brittle elements is shown in figure 7.10. If an element in such a system fails, its load-bearing capacity is completely exhausted. The other elements may or may not be able to prevent the system from failing completely, by redistribution of the loads. In real structures with a low degree of statical indeterminacy, the brittle failure of one element will usually result in the subsequent failure of other elements because of this redistribution. If this is the case, the system behaviour is like a series system. However, for structures with a high degree of statical indeterminacy and a relatively high safety factor the system in figure 7.10 may be a reasonable model because in such situations there will often be enough reserve strength capacity to carry the load after a brittle element failure.

Let $r_1, r_2, \ldots, r_n$, where $r_1 < r_2 < \ldots < r_n$, be the strength of the n elements shown in figure 7.10. The strength r of the system is then given by

$$r = \max(nr_1, (n-1)r_2, \ldots, 2r_{n-1}, r_n) \tag{7.8}$$

It can be shown that under certain conditions r is a realisation of a random variable R which for large n approaches a normal distribution.

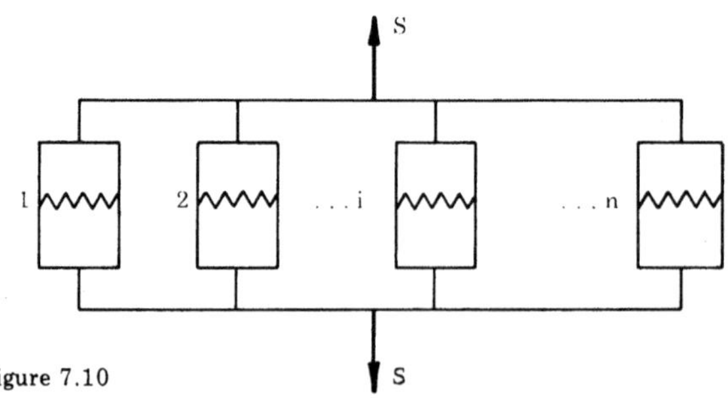

Figure 7.10

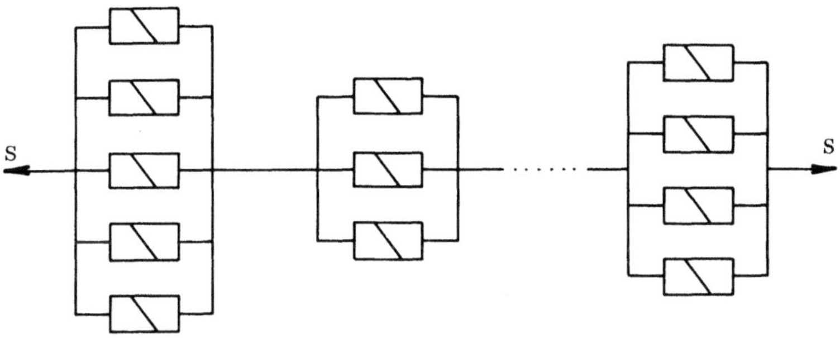

Figure 7.11

As mentioned above, failure of a statically indeterminate structure can sometimes be evaluated on the basis of a number of failure modes where each failure mode is modelled by a parallel system. Modelling of the complete structure will then be a series system of parallel subsystems as shown in figure 7.11. Note that a given element may appear in several failure modes.

Also note that correlation in such a system can appear at least in two forms, namely by correlation between single elements and correlation between failure modes. Systems with correlated elements or correlated subsystems will be treated in some detail in chapter 8. In the next section, fundamental systems with equally correlated elements will be discussed.

Example 7.4. Consider the statically indeterminate truss with 3 panels shown in figure 7.12. Assume that only the diagonals 1, 2, . . . , 6 can fail. This structure can then be modelled by the system shown in figure 7.13.

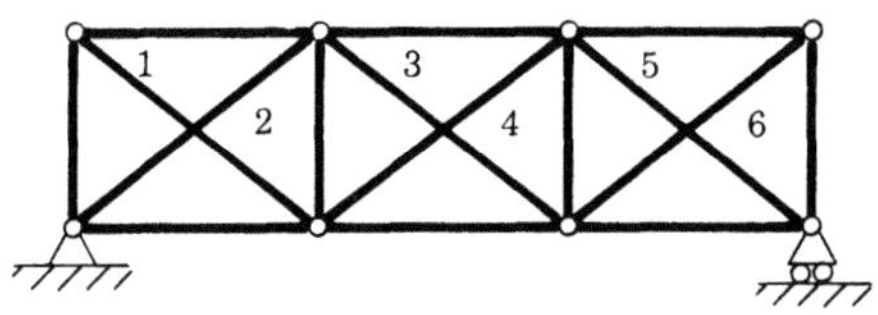

Figure 7.12

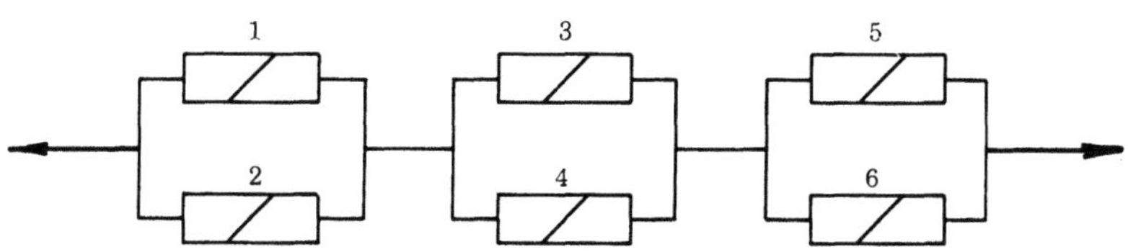

Figure 7.13

7.4 SYSTEMS WITH EQUALLY CORRELATED ELEMENTS

Exact calculation of the probability of failure for a given system with correlated strengths is usually not possible. However, bounds for the probability of failure can often be determined. This will be shown in chapter 8. However, in some important cases it is possible to calculate the exact probability of failure for the fundamental systems introduced in section 7.3. This has been done by Grigoriu & Turkstra for series systems and for parallel systems with ductile elements, on the assumption that the strength of the elements can be modelled by normally distributed random variables, R_i, i = 1, 2, . . . , n, which are equally correlated with a common correlation coefficient ρ. Further, it is assumed that the loads are deterministic and constant in time and all elements are designed in such a way that they have the same reliability index β_e. Let the strength R_i of element i be $N(\mu_i, \sigma_i)$ and S_i the load effect. Then

$$\beta_e = \frac{\mu_i - S_i}{\sigma_i}$$

or

$$\mu_i = S_i + \beta_e \sigma_i \tag{7.9}$$

If the coefficient of variation $V_i = \sigma_i/\mu_i$ is inserted into (7.9) this equation can be rewritten

$$\mu_i = S_i/(1 - \beta_e V_i) \tag{7.10}$$

The assumption of euqally correlated elements is relevant for some structures, but for a great number of structures such an assumption cannot be justified. However, it is of great advantage to use this assumption because the exact probability can then be calculated. Bearing this in mind it seems worthwhile investigating the possibility of using a kind of »average» coefficient of correlation in the general case where the correlation is unequal. Such an investigation has been performed for parallel systems with ductile elements and for series systems, and it has resulted in two new methods for calculating approximate values for the probability of failure for such systems (see chapter 8). In this section a brief presentation of the work by Grigoriu & Turkstra will be given.

For a *series system* with n elements it has been shown by Stuart that the probability of failure on the assumptions mentioned above is given by

$$P_f(\rho) = 1 - \int_{-\infty}^{\infty} [\Phi(\frac{\beta_e + \sqrt{\rho}\, t}{\sqrt{1-\rho}})]^n \varphi(t) dt \tag{7.11}$$

where Φ and φ denote the distribution and density function for the standard Gaussian random variable. The variation of the probability of failure P_f with ρ is shown in figure 7.14 for n = 1, 2, 5, and 10 and β_e = 3.0.

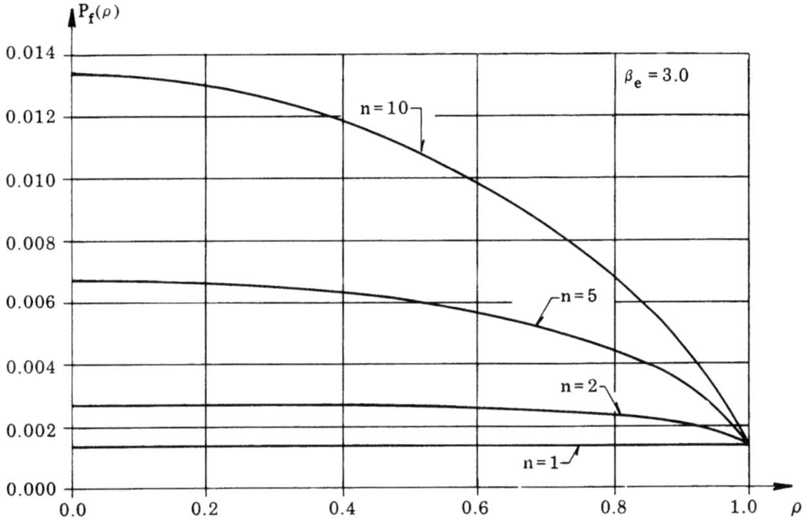

Figure 7.14

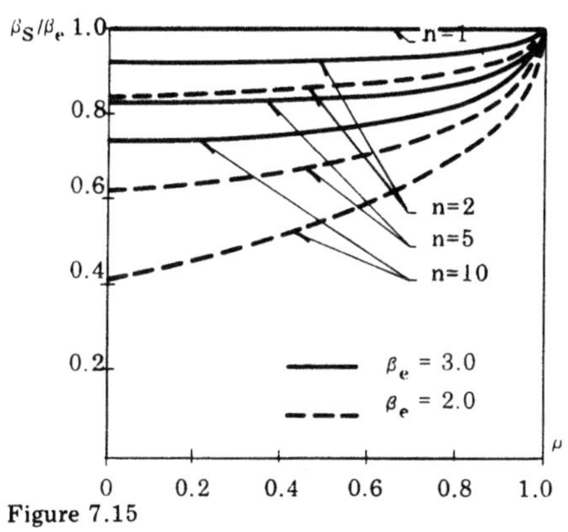

Figure 7.15

Note that, as expected, the probability of failure P_f decreases with the correlation coefficient ρ and increases with the number of elements n.

A formal reliability index β_S for the series system can now be calculated by

$$P_f = \Phi(-\beta_S) \iff \beta_S = -\Phi^{-1}(P_f) \tag{7.12}$$

The dependence of β_S/β_e on the correlation coefficient ρ for β_e = 3.0 and β_e = 2.0 and for n = 1, 2, 5, and 10 is shown in figure 7.15 (taken from the paper by Grigoriu & Turkstra).

It is seen from figures 7.14 and 7.15 that the reliability of a series system is increased if the dependence between the strength of the elements is increased. This result could be used in some design situations - for example, by stipulating that all members of a steel structure are from the same batch.

Next consider a *parallel system with n ductile elements* having normally distributed strengths R_i identically distributed $N(\mu, \sigma)$ and satisfying the same assumptions as above. According to equation (7.5), the strength R of the parallel system is equal to $\sum_{i=1}^{n} R_i$, with

$$E[R] = \sum_{i=1}^{n} E[R_i] = n\mu \tag{7.13}$$

and

$$\mathrm{Var}[R] = \sum_{i=1}^{n} \mathrm{Var}[R_i] + \rho \sum_{\substack{i,j=1 \\ i \neq j}}^{n} (\mathrm{Var}[R_i]\,\mathrm{Var}[R_j])^{\frac{1}{2}} = n\sigma^2 + n(n-1)\rho\sigma^2 \tag{7.14}$$

where ρ is the common correlation coefficient. Let the load on the system be S and the common element reliability index β_e. Then, according to equation (7.9)

$$S = \sum_{i=1}^{n} S_i = n\mu - n\beta_e\sigma \tag{7.15}$$

and the reliability index β_S for the system is

$$\beta_S = \frac{E[R] - S}{(\mathrm{Var}[R])^{\frac{1}{2}}} = \frac{n\mu - (n\mu - n\beta_e\sigma)}{(n\sigma^2 + n(n-1)\sigma^2\rho)^{\frac{1}{2}}} = \beta_e \sqrt{\frac{n}{1 + \rho(n-1)}} \tag{7.16}$$

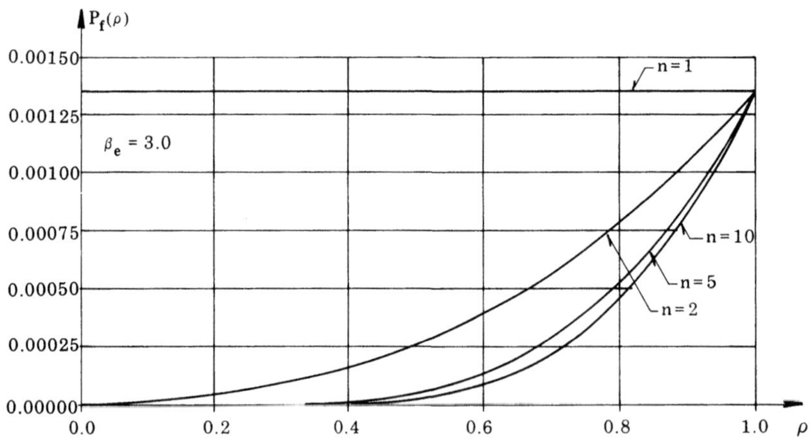

Figure 7.16

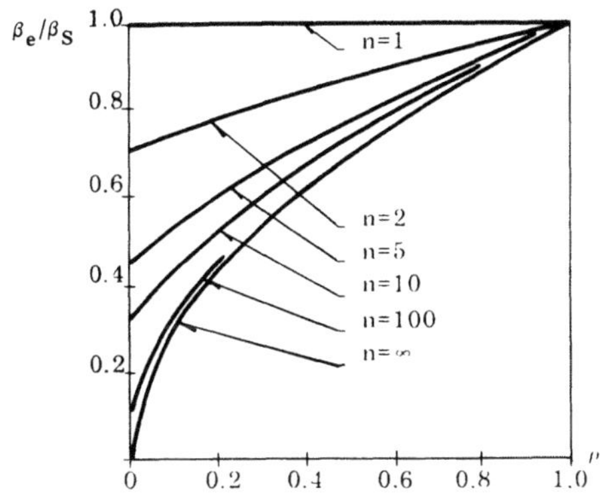

Figure 7.17

The variation of the probability of failure $P_f = \Phi(-\beta_S)$ is shown in figure 7.16 for n = 1, 2, 5, and 10 and $\beta_e = 3.0$. Note that P_f increases with the correlation coefficient ρ and decreases with the number of elements ($n > 1$). The dependence between the ratio β_e/β_S and ρ is shown in figure 7.17 for n = 1, 2, 5, 10, 100, and ∞. As expected β_S/β_e increases when ρ is decreased or n is increased. Therefore, the probability of failure for such a parallel system with correlated elements is underestimated if independence is assumed. This is in contrast to a series system.

Example 7.5. Consider again the structure in figure 7.12 modelled by the system shown in figure 7.13. With the same assumptions as above and the further assumptions that all diagonals are equally loaded and behave in a ductile manner, the probability of failure for the system can be calculated.

The reliability index β_p for a single panel with two diagonals is, in accordance with equation (7.16), given by

$$\beta_p = \beta_e \sqrt{2/(1+\rho)} \tag{7.17}$$

The probability of failure P_f for the system can now be calculated from (7.11) with n = 3, and β_e replaced by β_p, and ρ replaced by ρ_p, i.e. the correlation coefficient between the strength of the panels.

The correlation coefficient ρ_p can be determined in the following way. Let R_1 and R_2 be the strengths of the diagonals in a panel with the strength $R_p = R_1 + R_2$. Then $E[R_p] = 2\mu$, $\mathrm{Var}[R_p] = 2\sigma^2 + 2\rho\sigma^2 = 2(1+\rho)\sigma^2$ and

$$\mathrm{Cov}[R_1, R_2] = E[R_1 R_2] - E[R_1]E[R_2]$$

or

$$\rho\sigma^2 = E[R_1 R_2] - \mu^2 \tag{7.18}$$

Let R_p^1 and R_p^2 be the strengths of two panels. Then

$$E[R_p^1 R_p^2] = 4E[R_1 R_2] = 4(\rho\sigma^2 + \mu^2) \tag{7.19}$$

: Finally

$$\rho_p = \frac{E[R_p^1 R_p^2] - E[R_p]^2}{Var[R_p]} = \frac{4(\rho\sigma^2 + \mu^2) - 4\mu^2}{2(1+\rho)\sigma^2} = \frac{2\rho}{1+\rho} \tag{7.20}$$

: The probability of failure P_f for the system is therefore

$$P_f = 1 - \int_{-\infty}^{\infty} [\Phi(\frac{\beta_p + \sqrt{\rho_p}\, t}{\sqrt{1-\rho_p}})]^3 \varphi(t)dt \tag{7.21}$$

: where β_p is given by (7.17) and ρ_p by (7.20).

It is obvious from example 7.5 that the only calculation problem in estimating the probability of failure for structural systems satisfying the assumptions mentioned above is the integral in equation (7.11). For β_e = 3.00 values of P_f are tabulated in Table 7.1 for n = 1, 2, . . . , 10 and $\rho \in [0 ; 1.00]$. These values have been calculated by numerical integration using Simpson's formula and 200 intervals.

	n									
ρ	1	2	3	4	5	6	7	8	9	10
0.00	13.50	26.98	40.44	53.89	67.31	80.72	94.11	107.4	120.8	134.2
0.05	13.50	26.97	40.41	53.81	67.19	80.54	93.86	107.1	120.4	133.6
0.10	13.50	26.95	40.35	53.70	67.01	80.27	93.48	106.6	119.8	132.8
0.15	13.50	26.92	40.27	53.74	66.75	79.88	92.93	105.9	118.8	131.7
0.20	13.50	26.88	40.16	53.32	66.38	79.34	92.19	104.9	117.6	130.2
0.25	13.50	26.83	40.00	53.02	65.88	78.61	91.20	103.6	116.0	128.2
0.30	13.50	26.76	39.80	52.62	65.24	77.68	89.93	102.0	113.9	125.7
0.35	13.50	26.67	39.53	52.10	64.42	76.49	88.34	99.96	111.4	122.6
0.40	13.50	26.54	39.18	51.45	63.39	75.03	86.39	97.49	108.3	119.0
0.45	13.50	26.38	38.74	50.64	62.13	73.26	84.05	94.55	104.8	114.7
0.50	13.50	26.18	38.19	49.64	60.61	71.14	81.13	91.12	100.6	109.9
0.55	13.50	25.92	37.52	48.44	58.79	68.66	78.11	87.19	95.93	104.4
0.60	13.50	25.60	36.69	46.99	56.66	65.78	74.45	82.72	90.65	98.26
0.65	13.50	25.20	35.69	45.28	54.16	62.47	70.30	77.71	84.76	91.50
0.70	13.50	24.70	34.48	43.26	51.28	58.69	65.61	72.11	78.25	84.09
0.75	13.50	24.07	33.01	40.87	47.94	54.39	60.34	65.89	71.10	76.01
0.80	13.50	23.28	31.24	38.05	44.07	49.49	54.44	59.00	63.47	67.23
0.85	13.50	22.26	29.05	34.69	39.56	43.89	47.77	51.32	54.59	57.63
0.90	13.50	20.89	26.27	30.58	34.19	37.33	40.11	42.62	44.90	47.00
0.95	13.50	18.91	22.49	25.20	27.40	29.26	30.87	32.29	33.56	34.72
1.00	13.50	13.50	13.50	13.50	13.50	13.50	13.50	13.50	13.50	13.50

Table 7.1. $P_f \times 10^4$ after (7.11) with β_e = 3.00

Example 7.6. Consider the same structure as in examples 7.4 and 7.5 and let $\rho = 0.8$ and $\beta_e = 2.85$. By equation (7.17) the reliability index β_p for the panels is

$$\beta_p = \beta_e \sqrt{2/(1+\rho)} = 2.85 \cdot \sqrt{2/1.8} = 3.00$$

and the correlation coefficient ρ_p between the strength of the panels is

$$\rho_p = 2\rho/(1+\rho) = 1.6/1.8 = 0.889$$

The probability of failure P_f for the system is then (use table 7.1 with n = 3)

$$P_f = 26.88 \cdot 10^{-4} \tag{7.22}$$

Exercise 7.3. Consider the statically determinate structure shown in figure 7.18. Assume that this structure can be modelled by a series system with 7 elements satisfying all the assumptions mentioned above. Let the reliability index β_e for the elements be $\beta_e = 3.00$ and the correlation coefficient between the elements $\rho = 0.85$. Calculate the probability of failure P_f for this system and compare P_f with the probability of failure for a single element.

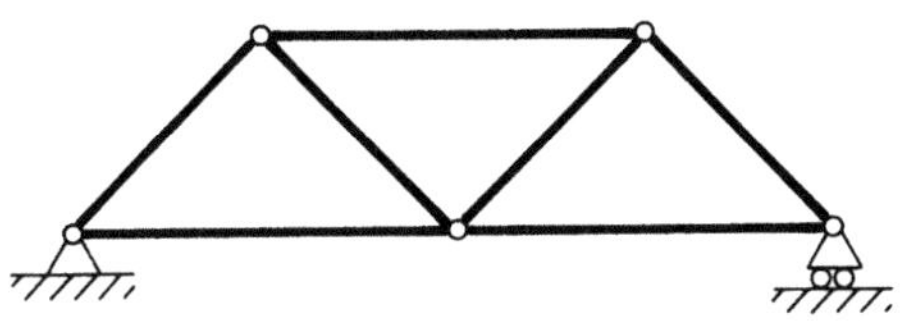

Figure 7.18

BIBLIOGRAPHY

[7.1] Daniels, H. E.: *The Statistical Theory of the Strength of Bundles of Threads.* Royal Stat. Soc., Series A, Vol. 133, 1945, p. 405.

[7.2] Grigoriu, M. & C. Turkstra: *Safety of Structural Systems with Correlated Resistances.* Applied Math. Modelling, Vol. 3, 1979, pp. 130-136.

[7.3] Stuart, A. J.: *Equally Correlated Variates and the Multinormal Integral.* J. Royal Stat. Soc., Series B, Vol. 20, 1958, pp. 373-378.

[7.4] Thoft-Christensen, P.: *Fundamentals of Structural Reliability.* Lectures on Structural Reliability (ed. P. Thoft-Christensen), Aalborg University Centre, Aalborg, Denmark, 1980, pp. 1-28.

[7.5] Thoft-Christensen, P. & J. D. Sørensen: *Reliability of Structural Systems with Correlated Elements.* Applied Mathematical Modelling, Vol. 6, 1982.

Chapter 8

RELIABILITY BOUNDS FOR STRUCTURAL SYSTEMS

8.1 INTRODUCTION

In chapter 7, the concept of modelling of structural systems by series and parallel systems was introduced. It was shown that in general the exact determination of the probability of failure of such systems is not possible and that a numerical calculation is often rather time-consuming. However, upper and lower bounds for the exact probability of failure can often be formulated, but the practical value of such bounds depends on how narrow they are. In this chapter two sets of bounds will be derived, namely the so-called *simple bounds* and *Ditlevsen bounds*.

For systems with equally correlated elements, the probability of failure can be calculated exactly for a series system with n elements. This can have some important practical applications. It is shown in section 7.4 that, under these assumptions, the probability of failure is given by (7.11)

$$P_f(\rho) = 1 - \int_{-\infty}^{\infty} [\Phi(\frac{\beta_e + \sqrt{\rho}\, t}{\sqrt{1-\rho}})]^n \varphi(t)dt \qquad (7.11)$$

where β_e is the common reliability index for all n elements and ρ the common correlation coefficient between any pair of elements.

In the same section 7.4, the following relation between the reliability index β_S for a parallel system and the common reliability index β_e for the n ductile elements was given as

$$\beta_S = \beta_e \sqrt{\frac{n}{1 + \rho(n-1)}} \qquad (7.16)$$

The assumptions under which (7.16) is valid are specified in section 7.4.

For a system with unequal correlation coefficients it is now natural to investigate the approximation achieved by using the average correlation coefficient $\bar{\rho}$ in connection with equations (7.11) and (7.16). This has been investigated by Thoft-Christensen & Sørensen. They also suggest a very simple method to get approximate values for the probability of failure for a series system based on an equivalent correlation coefficient $\bar{\bar{\rho}}$. These approximate methods are presented in sections 8.4 and 8.5.

8.2 SIMPLE BOUNDS

If a system has a great number of elements it is convenient to use Boolean variables to describe the state of the elements. It is beyond the scope of this chapter to give a detailed introduction to Boolean algebra, but all that is needed here is to assume that each element can exist in only one of two states, namely »failure» or »non-failure», and to associate with element i, i = 1, 2, ... , n a so-called Boolean variable (indicator function) S_i defined by

$$S_i = \begin{cases} 1 & \text{if the element is in a non-failure state} \\ 0 & \text{if the element is in a failure state} \end{cases} \tag{8.1}$$

Further, it is useful to define a set of Boolean variables F_i by

$$F_i = 1 - S_i = \begin{cases} 0 & \text{if the element is in a non-failure state} \\ 1 & \text{if the element is in a failure state} \end{cases} \tag{8.2}$$

Consider a system with n elements. By associating the state of each element by a variable S_i, i = 1, 2, . . . , n, the state of the system is determined by a vector

$$\bar{S} = (S_1, S_2, \ldots, S_n) \tag{8.3}$$

and can be described by a so-called *system function* $S_S(\bar{S})$. The system function itself is a Boolean variable defined by

$$S_S(\bar{S}) = \begin{cases} 1 & \text{if the system is in a non-failure state} \\ 0 & \text{if the system is in a failure state} \end{cases} \tag{8.4}$$

For a *series system* one gets by analogy with equation (7.3) that

$$S_S(\bar{S}) = S_1 S_2 \ldots S_n = \prod_{i=1}^{n} S_i \tag{8.5}$$

or (by definition)

$$S_S(\bar{F}) = 1 - S_S(\bar{S}) = 1 - \prod_{i=1}^{n} (1 - F_i) \tag{8.6}$$

Note that

$$S_S(\bar{F}) = \begin{cases} 0 & \text{if the system is in a non-failure state} \\ 1 & \text{if the system is in a failure state} \end{cases} \tag{8.7}$$

For a *parallel system* the system function is given by

$$S_S(\bar{S}) = 1 - \prod_{i=1}^{n} (1 - S_i) \tag{8.8}$$

or

$$S_S(\overline{F}) = \prod_{i=1}^{n} F_i \tag{8.9}$$

Return now to a *single element* i. From the definition of S_i and F_i it follows that the expected values $E[S_i]$ and $E[F_i]$ are associated with the probability of failure $P(F_i = 1)$ for element i, in the following way

$$E[S_i] = 1 \cdot P(S_i = 1) + 0 \cdot P(S_i = 0) = P(F_i = 0) = 1 - P(F_i = 1) \tag{8.10}$$

and

$$E[F_i] = 1 \cdot P(F_i = 1) + 0 \cdot P(F_i = 0) = P(F_i = 1) \tag{8.11}$$

The probability of failure P_f for the *system* is then by analogy

$$P_f = P(S_S(\overline{F}) = 1) = E[S_S(\overline{F})] \tag{8.12}$$

As mentioned earlier, determination of the probability of failure P_f for a system on the basis of equation (8.12) implies the calculation of n-dimensional multi-integrals. However, very simple bounds for P_f can be derived for a *series system* with positive correlation between the strength of the elements.

Assume that

$$P(\bigcap_{j=1}^{i+1} S_j = 1) \geqslant P(S_{i+1} = 1)\, P(\bigcap_{j=1}^{i} S_j = 1) \tag{8.13}$$

for all $1 \leqslant i \leqslant n-1$. The condition (8.13) expresses that the probability of non-failure for the system consisting of the elements $j = 1, 2, \ldots, i+1$ is greater than or equal to the product of the probability of non-failure for the system consisting of the elements $j = 1, 2, \ldots, i$ and the probability of non-failure for element $i+1$ (see figure 8.1).

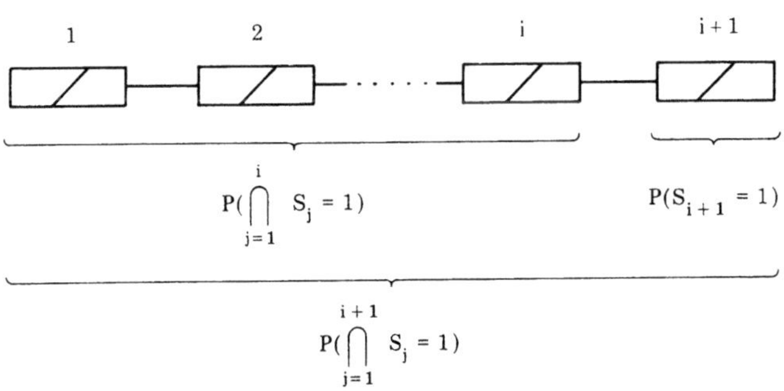

Figure 8.1

From (8.13) follows that

$$P_f = P(\bigcup_{i=1}^{n} F_i = 1) = 1 - P(\bigcap_{i=1}^{n} S_i = 1) =$$

$$1 - P(S_1 = 1)P(S_2 = 1 | S_1 = 1) \dots P(S_n = 1 | S_1 = 1 \cap \dots \cap S_{n-1} = 1) \leq$$

$$1 - P(S_1 = 1)P(S_2 = 1) \dots P(S_n = 1) = 1 - \prod_{i=1}^{n} (1 - P(F_i = 1)) \tag{8.14}$$

Clearly

$$P_f \geq \max_{i=1,n} P(F_i = 1) \tag{8.15}$$

The upper and lower bounds for series systems (8.14) and (8.15) can be combined

$$\max_{i=1,n}^{n} P(F_i = 1) \leq P_f \leq 1 - \prod_{i=1}^{n} (1 - P(F_i = 1)) \tag{8.16}$$

Example 8.1. Consider the series system with two elements shown in figure 8.2 and assume that the reliability index $\beta_e = 3.00$. Further assume that the equation (7.11) can be used for estimating the probability of failure. The probability of failure for each element is

$$P(F_1 = 1) = P(F_2 = 1) = \Phi(-3.00) = 0.00135 \tag{8.17}$$

Bounds for the probability of failure P_f for the system are according to (8.16)

$$0.00135 \leq P_f \leq 1 - (1 - 0.00135)^2$$

or

$$0.00135 \leq P_f \leq 0.00270$$

and are in good agreement with table 7.1.

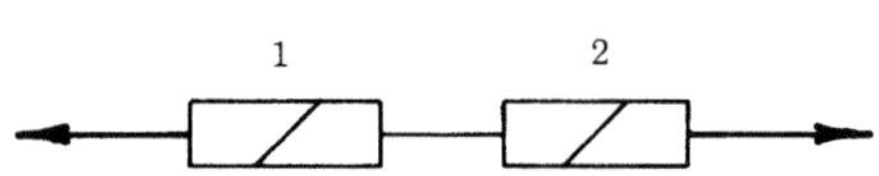

Figure 8.2

The lower bound in (8.16) is equal to the exact value for P_f if there is perfect dependence between all elements (the correlation coefficient between any pair of elements is equal to 1). The upper bound in (8.16) corresponds to no dependence between any pair of elements (see equation (7.3)).

For a *parallel system* a simple lower bound and a simple upper bound can be constructed by arguing that perfect correlation between all elements now corresponds to the upper bound and no correlation between any pair of elements corresponds to the lower bound. Therefore,

$$\prod_{i=1}^{n} P(F_i = 1) \leq P_f \leq \min_{i=1,n} P(F_i = 1) \tag{8.18}$$

: **Example 8.2.** Consider one of the panels from the structural system shown in fig. 7.12. This panel is modelled in figure 7.13 by a parallel system with only two elements. In example 7.5 and 7.6 the reliability index β_p for such a panel is calculated assuming that the reliability index for a single element is $\beta_e = 2.85$ and that the correlation coefficient between the elements is $\rho = 0.8$. One gets $\beta_p = 3.00$. The probability of failure for each element is

$$P(F_1 = 1) = P(F_2 = 1) = \Phi(-2.85) = 0.00219$$

Therefore, by (8.18)

$$0.00219^2 \leq P_f \leq 0.00219$$

or

$$4.8 \cdot 10^{-6} \leq P_f \leq 0.219 \cdot 10^{-2} \tag{8.19}$$

for $0 \leq \rho \leq 1$. In the particular case with $\rho = 0.8$ one gets

$$P_f = \Phi(-3.00) = 0.135 \cdot 10^{-2}$$

8.3 DITLEVSEN BOUNDS

The simple bounds presented in chapter 8.2 are usually rather wide because they correspond, respectively, to perfect dependence between all elements and no dependence between any pair of elements. Several other bounds have been suggested in the literature. In this chapter a brief derivation will be given for the bounds by Ditlevsen for series systems. These bounds are very narrow, especially for correlation coefficients below 0.6.

It follows from equation (8.6) that

$$\begin{aligned} S_S(\overline{F}) &= 1 - S_1 S_2 \ldots S_n \\ &= 1 - S_1 S_2 \ldots S_{n-1} + S_1 S_2 \ldots S_{n-1} F_n \\ &= \ldots \\ &= F_1 + S_1 F_2 + S_1 S_2 F_3 + \ldots + S_1 S_2 \ldots S_{n-1} F_n \end{aligned} \tag{8.20}$$

Hence, in accordance with (8.12), the probability of failure P_f is

$$P_f = E[F_1] + E[S_1 F_2] + \ldots + E[S_1 S_2 \ldots S_{n-1} F_n] \tag{8.21}$$

It is easy to see that

$$1 - (F_1 + F_2 + \ldots + F_{n-1}) \leqslant S_1 S_2 \ldots S_{n-1} \leqslant 1 - F_i \tag{8.22}$$

for i = 1, 2, 3, . . . , n − 1. By inserting (8.22) into (8.21) and bearing in mind that all probabilities are non-negative one gets the following upper and lower bounds for a series system

$$P_f \leqslant \sum_{i=1}^{n} P(F_i = 1) - \sum_{i=2}^{n} \max_{j<i} P((F_i = 1) \cap (F_j = 1)) \tag{8.23}$$

and

$$P_f \geqslant P(F_1 = 1) + \sum_{i=2}^{n} \max[P(F_i = 1) - \sum_{j=1}^{i-1} P((F_i = 1) \cap (F_j = 1)),\ 0] \tag{8.24}$$

The numbering of the elements may influence the bounds (8.23) and (8.24). Therefore, to get the best bounds one has to choose from the different possible numberings of the elements.

In section 8.5, the bounds (8.23) and (8.24) will be used to determine approximate methods to estimate the probability of failure of series systems.

8.4 PARALLEL SYSTEMS WITH UNEQUALLY CORRELATED ELEMENTS

In a real structure modelled by a parallel system with n elements, the correlation coefficient between the elements will usually not be equal. Very simple bounds for the probability of failure for such a system are presented in section 8.2, equation (8.18), valid when the correlation between any pair of elements is positive. Unfortunately, these bounds are rather wide because they correspond to equal correlation between any pair of elements, namely $\rho = 1$ for the upper bound and $\rho = 0$ for the lower bound.

In chapter 7.4 parallel systems with n ductile elements were investigated under the following assumptions

(1) the loading is deterministic and constant in time

(2) the strength R_i, i = 1, 2, . . . , n of the members is identically normally distributed $N(\mu, \sigma)$

(3) all elements are designed to have a common reliability index β_e

(4) common correlation coefficient ρ between any pair of elements.

Under these assumptions it was shown that the reliability index β_S for the parallel system is related to the reliability index β_e for the elements by (7.16)

$$\beta_S = \beta_e \sqrt{\frac{n}{1+\rho(n-1)}} \tag{7.16}$$

Now the assumption (4) above will be relaxed. The correlation coefficient between element i and j will be denoted ρ_{ij} and the corresponding *correlation matrix* $\bar{\bar{C}}$ is defined by

$$\bar{\bar{C}} = [\rho_{ij}] = \begin{bmatrix} 1 & \rho_{12} \cdots\cdots & \rho_{1n} \\ \rho_{21} & 1 \cdots\cdots & \rho_{2n} \\ \vdots & \ddots & \vdots \\ \rho_{n1} & \rho_{n2} \cdots\cdots & 1 \end{bmatrix} \tag{8.25}$$

The reliability index β_S for such a parallel system can now be calculated in a similar way as used in deriving (7.16) in section 7.4. One gets

$$\beta_S = \frac{E[R]-S}{(Var[R])^{\frac{1}{2}}} = (n\mu - (n\mu - n\beta_e\sigma))(n\sigma^2 + \sigma^2 \sum_{\substack{i,j=1\\ i\neq j}}^{n} \rho_{ij})^{-\frac{1}{2}} = \beta_e \sqrt{\frac{n}{1+\bar{\rho}(n-1)}} \tag{8.26}$$

where

$$\bar{\rho} = \frac{1}{n(n-1)} \sum_{\substack{i,j=1\\ i\neq j}}^{n} \rho_{ij} \tag{8.27}$$

By comparing (7.16) and (8.26) it is seen that for systems with non-equal correlation coefficients the reliability index β_S can be calculated by the simple expression (7.16) simply by inserting for ρ the *average correlation coefficient* $\bar{\rho}$ defined by (8.27). $\bar{\rho}$ is the average of all ρ_{ij}, $i \neq j$.

This result is interesting because calculation of the probability of failure for such a system can easily now be made for any correlation matrix (8.25). One only needs to make a set of curves, as shown in figure 7.16, once.

Example 8.3. Consider a parallel system, with 5 elements, as shown in figure 8.3.

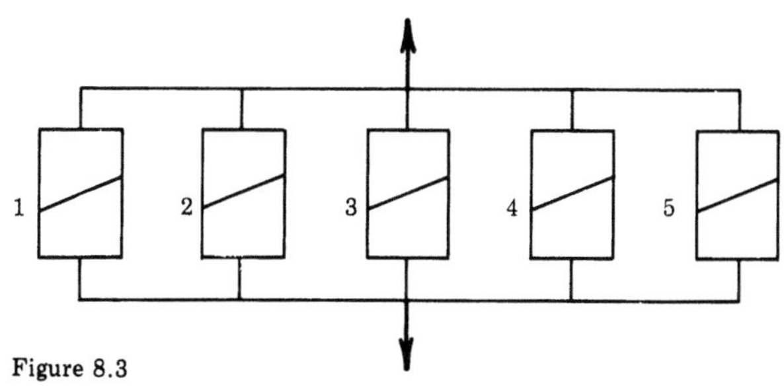

Figure 8.3

: Assume that the assumptions (1), (2) and (3) above are fulfilled, with $\beta_e = 3.50$ and let
: the correlation matrix be

$$\bar{\bar{C}} = \begin{bmatrix} 1 & 0.5 & 0.2 & 0.1 & 0 \\ 0.5 & 1 & 0.5 & 0.2 & 0.1 \\ 0.2 & 0.5 & 1 & 0.5 & 0.2 \\ 0.1 & 0.2 & 0.5 & 1 & 0.5 \\ 0 & 0.1 & 0.2 & 0.5 & 1 \end{bmatrix}$$

: The average correlation coefficient is

$$\bar{\rho} = \frac{5.6}{5 \cdot 4} = 0.28$$

: The reliability index for the system is therefore

$$\beta_S = 3.50 \sqrt{\frac{5}{1 + 0.28 \cdot 4}} = 5.38 \qquad (8.28)$$

: **Exercise 8.1.** Compare the exact result (8.28) with the approximate results obtained by
: assuming no correlation $\rho_{ij} = 0$ and perfect correlation $\rho_{ij} = 1$.

8.5 SERIES SYSTEMS WITH UNEQUALLY CORRELATED ELEMENTS

In this section series systems with unequal correlation between their elements are considered. An exact determination of the probability of failure P_f for such a system will, as mentioned earlier, imply extensive numerical calculations of multi-integrals. In section 8.2 the following very simple bounds were derived for a series system with n elements

$$\max_{i=1,n} P(F_i = 1) \leq P_f \leq 1 - \prod_{i=1}^{n} (1 - P(F_i = 1)) \qquad (8.16)$$

where $P(F_i = 1)$ is the probability of failure for element i. Much better upper and lower bounds are the Ditlevsen bounds (8.23) and (8.24).

In some special cases, the exact probability of failure P_f can be calculated relatively easily. This has been shown in section 7.4 under the following assumptions:

(1) the loading is deterministic and constant in time

(2) the strength R_i, $i = 1, 2, \ldots, n$ of the members is identically normally distributed $N(\mu, \sigma)$

(3) all elements are designed to have a common reliability index β_e

(4) common correlation coefficient ρ between any pair of elements.

Under these assumptions the probability of failure P_f is given by (7.11)

$$P_f = 1 - \int_{-\infty}^{\infty} [\Phi(\frac{\beta_e + \sqrt{\rho}\, t}{\sqrt{1-\rho}})]^n \varphi(t) dt \qquad (7.11)$$

Assumption (4) will now be relaxed. In section 8.4 it is shown that use of the average correlation coefficient $\bar{\rho}$, defined by (8.27), gives the correct value of P_f for a parallel system with ductile elements. It is therefore natural to investigate the use of $\bar{\rho}$ as an equivalent correlation coefficient and then use equation (7.11). Note that $\bar{\rho} = \rho$ when the assumption (4) is valid. The use of $\bar{\rho}$ has been tested with a series system, with the following simple correlation matrix

$$\bar{\bar{C}} = \begin{bmatrix} 1 & \rho_{12} & 0.2 & \cdots & 0.2 \\ \rho_{12} & 1 & 0.2 & \cdots & 0.2 \\ 0.2 & 0.2 & 1 & \cdots & 0.2 \\ \vdots & \vdots & \vdots & \ddots & \vdots \\ 0.2 & 0.2 & 0.2 & \cdots & 1 \end{bmatrix} \tag{8.29}$$

For this example approximate values for the probability of failure can be calculated from (7.11) with ρ equal to the average correlation coefficient $\bar{\rho}$ defined by (8.27). The result is shown in figure 8.4 for n = 2, 3, 5, and 10, and $\beta_e = 3.00$ as the curves (2).

The upper and lower bounds (8.23) and (8.24) by Ditlevsen have been calculated for some values of ρ_{12} and are shown in figure 8.4 as intervals (4). The curves (1) are upper bounds corresponding to independent elements (equation (8.16)). The curves (3) are approximations based on the so-called equivalent correlation coefficient $\bar{\bar{\rho}}$ defined later in this section.

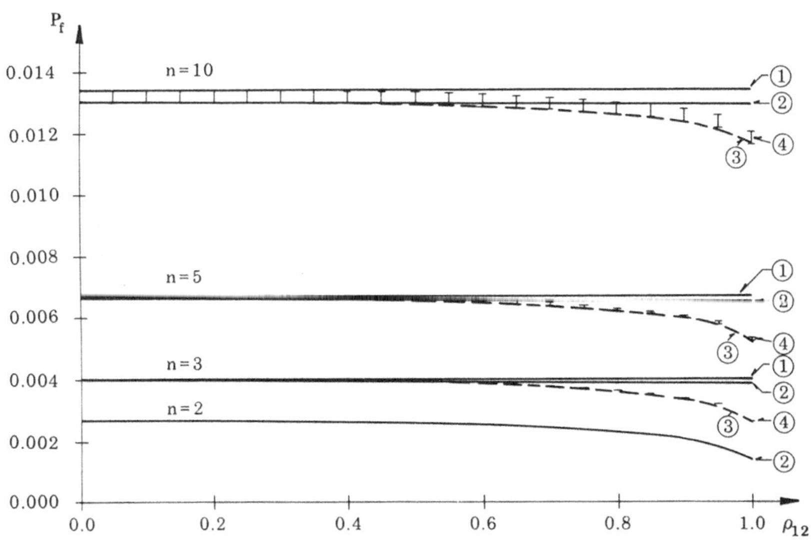

Figure 8.4

It appears from figure 8.4 that the Ditlevsen bounds (8.23) and (8.24) are very narrow for $n = 3$ and $n = 5$. For $n = 10$ some gap is disclosed. As P_f for $\rho_{12} = 0.2$ is calculated exactly using $\rho = \bar{\rho}$, the lower bound (8.24) is apparently very close to the exact value for P_f for ρ_{12} in the neighbourhood of 0.2.

For $\rho_{12} > 0.7$ it is clear from figure 8.4 that using $\bar{\rho}$ as an equivalent correlation coefficient is on the safe side. The curves (2) do not decrease as fast as the correct values of P_f for ρ_{12} approaching 1.0. To obtain a better agreement a modification of $P(\bar{\rho})$ must be used.

A much better agreement can be obtained by using an equivalent correlation coefficient $\bar{\bar{\rho}}$ indirectly defined by

$$P_f(\bar{\bar{\rho}}) = P_f(\bar{\rho}) - (P_{f,2}(\bar{\rho}) - P_{f,2}(\rho_{max})) \tag{8.30}$$

where $P_{f,2}(\rho_{max})$ is the probability of failure calculated from equation (7.11) with $n = 2$ and $\rho = \rho_{max}$, where

ρ	β						
	2.00	2.50	3.00	3.50	4.00	4.50	5.00
	$\times 10^{-3}$	$\times 10^{-4}$	$\times 10^{-4}$	$\times 10^{-5}$	$\times 10^{-6}$	$\times 10^{-7}$	$\times 10^{-8}$
0.00	44.98	123.81	26.98	46.52	63.34	67.95	57.33
0.05	44.82	123.63	26.97	46.52	63.34	67.95	57.33
0.10	44.63	123.39	26.95	46.51	63.34	67.95	57.33
0.15	44.40	123.08	26.92	46.49	63.33	67.95	57.33
0.20	44.13	122.68	26.88	46.47	63.32	67.95	57.33
0.25	43.82	122.18	26.83	46.43	63.30	67.94	57.33
0.30	43.46	121.57	26.76	46.37	63.27	67.93	57.32
0.35	43.05	120.82	26.66	46.29	63.23	67.91	57.32
0.40	42.58	119.90	26.54	46.18	63.15	67.88	57.31
0.45	42.05	118.81	26.38	46.01	63.03	67.82	57.29
0.50	41.45	117.50	26.18	45.79	62.86	67.72	57.25
0.55	40.77	115.95	25.92	45.48	62.59	67.56	57.17
0.60	40.00	114.10	25.60	45.07	62.22	67.30	57.05
0.65	39.13	111.92	25.20	44.53	61.67	66.90	56.83
0.70	38.14	109.34	24.70	43.81	60.90	66.29	56.48
0.75	37.00	106.25	24.07	42.85	59.82	65.37	55.88
0.80	35.68	102.51	23.28	41.58	58.27	63.96	54.91
0.85	34.10	97.91	22.26	39.86	56.07	61.80	53.31
0.90	32.14	92.01	20.89	37.45	52.80	58.41	50.61
0.95	29.48	83.73	18.91	33.77	47.54	52.61	45.68
1.00	22.75	62.10	13.50	23.26	31.67	33.98	28.67

Table 8.1. P_f after (7.11) with $n = 2$.

$$\rho_{max} = \max_{\substack{i,j=1,n \\ i \neq j}} \rho_{ij} \tag{8.31}$$

Note that $P_f(\bar{\bar{\rho}})$ gives the correct value when all ρ_{ij}, $i \neq j$, are equal. $P_f(\bar{\bar{\rho}})$ is shown on figure 8.4. as the curves (3) and it can be concluded that in this example the values of $P_f(\bar{\bar{\rho}})$ are close to the (lower) bounds, but a little on the unsafe side.

It is very convenient to use approximate values for P_f based on $P_f(\bar{\rho})$ or $P_f(\bar{\bar{\rho}})$, because one only needs a table giving P_f as a function of n and ρ defined by (7.11). Then, for any correlation matrix $\bar{\bar{C}}$, approximate values can easily be calculated by hand.

To facilitate the use of equation (8.30) $P_{f,2}$ is tabulated in table 8.1.

Example 8.4. Consider a series system with the same 5 elements as in example 8.3. Let $\beta_e = 3.00$ but let the correlation coefficients be unchanged from example 8.3.

The average correlation coefficient is

$$\bar{\rho} = 0.28$$

An approximate value P_f' for the probability of failure P_f is given by equation (7.11) (use table 7.1 or figure 8.4)

$$P_f' = 65.5 \cdot 10^{-4}$$

It can be shown that the Ditlevsen bounds are $63.8 \cdot 10^{-4} < P_f < 64.2 \cdot 10^{-4}$ so that using $\bar{\rho}$ is on the safe side in this case. The same conclusion is true when $\bar{\bar{\rho}}$ is used. One gets by (8.30) that (use table 7.1 or table 8.1)

$$P_f'' = P_f(\bar{\rho}) - (P_{f,2}(\bar{\rho}) - P_{f,2}(\rho_{max})) = P_f(0.28) - (P_{f,2}(0.28) - P_{f,2}(0.5))$$

$$= 65.5 \cdot 10^{-4} - (26.8 \cdot 10^{-4} - 26.2 \cdot 10^{-4}) = 64.9 \cdot 10^{-4}$$

Example 8.5. Consider a series system with 10 ductile elements. Assume that the assumptions (1),(2) and (3) on page 136 are fulfilled, with $\beta_e = 3.00$ and let $\bar{\rho} = 0.8$ and $\rho_{max} = 0.9$. From equation (7.11) (use table 7.1) the following approximate value P_f' for the probability of failure can be determined

$$P_f' = 67.23 \cdot 10^{-4}$$

The approximate value P_f'' based on the equivalent correlation coefficient $\bar{\bar{\rho}}$ is then (use table 8.1)

$$P_f'' = P_f(0.8) - (P_{f,2}(0.8) - P_{f,2}(0.9))$$

$$= 67.2 \cdot 10^{-4} - (23.3 \cdot 10^{-4} - 20.9 \cdot 10^{-4}) = 64.8 \cdot 10^{-4}$$

: The simple bounds for this example can be calculated from (8.16) with $P(F_i=1)=13.5\cdot10^{-4}$
: for i = 1, 2, . . . , 10. One gets
:
: $$13.5\cdot10^{-4}\leqslant P_f\leqslant 1-(1-13.5\cdot10^{-4})^{10}$$
:
: $$13.5\cdot10^{-4}\leqslant P_f\leqslant 134.2\cdot10^{-4}$$

The approximations $P_f(\bar{\rho})$ and $P_f(\bar{\bar{\rho}})$ have been evaluated by Thoft-Christensen & Sørensen by extensive simulations. A great number of correlation matrices $\bar{\bar{C}}$ have been generated randomly assuming the single elements of $\bar{\bar{C}}$ to be uniformly distributed in the interval [0;1]. Non-positive definite matrices were of course rejected in this testing. The results of these simulations have been compared with the Ditlevsen bounds and it can be concluded that using the equivalent correlation coefficient gives approximate values close to these bounds. Unfortunately it is not possible to conclude that the use of $P_f(\bar{\rho})$ and $P_f(\bar{\bar{\rho}})$ is on the safe side.

A final illustration of the different methods available to evaluate the probability of failure of series systems is made for a system with the following correlation matrix

$$\bar{\bar{C}}=\begin{bmatrix}1 & \rho & \rho^2 & \cdots & \rho^n\\ \rho & 1 & \rho & \cdots & \rho^{n-1}\\ \rho^2 & \rho & 1 & \cdots & \rho^{n-2}\\ \vdots & \vdots & \vdots & \ddots & \vdots\\ \rho^n & \rho^{n-1} & \rho^{n-2} & \cdots & 1\end{bmatrix} \qquad (8.32)$$

The results for $\beta_e = 3.00$ are shown in figure 8.5 for n = 3, 5 and 10. The upper bounds $P_f(0)$ are the curves (1) . $P_f(\bar{\rho})$ are the curves (2) and $P_f(\bar{\bar{\rho}})$ the curves (3) . The upper and lower bounds by Ditlevsen are shown as intervals (4) . In this case the curves for $P_f(\bar{\rho})$ and $P_f(\bar{\bar{\rho}})$ are rather close and they differ only slightly from the upper bounds. Note the great improvement obtained by using $P_f(\bar{\rho})$ or $P_f(\bar{\bar{\rho}})$ instead of $P_f(0)$, where $P_f(0)$ corresponds to assuming independent elements.

With regard to the Ditlevsen bounds, it can be concluded from figure 8.5 that they may be rather wide, especially for correlation coefficients greater than 0.7 and when n increases. For small correlation coefficients the bounds are narrow and are therefore of great importance in evaluating the probability of failure for series systems.

For a statically indeterminate structural system, a failure mode can often be modelled by a parallel system. Usually several failure modes exist and the total modelling will be a series of parallel systems as shown in fig. 7.11. In using the approximate methods presented above for series systems, knowledge of the reliability index β for each parallel system is required and, further, the correlation between the parallel systems must be known. Note that the methods above are based on the assumption that all parallel systems have the same reliability index. This will usually not be the case. To overcome the difficulty it may be useful to use an *equivalent reliability index* $\bar{\bar{\beta}}$ defined by

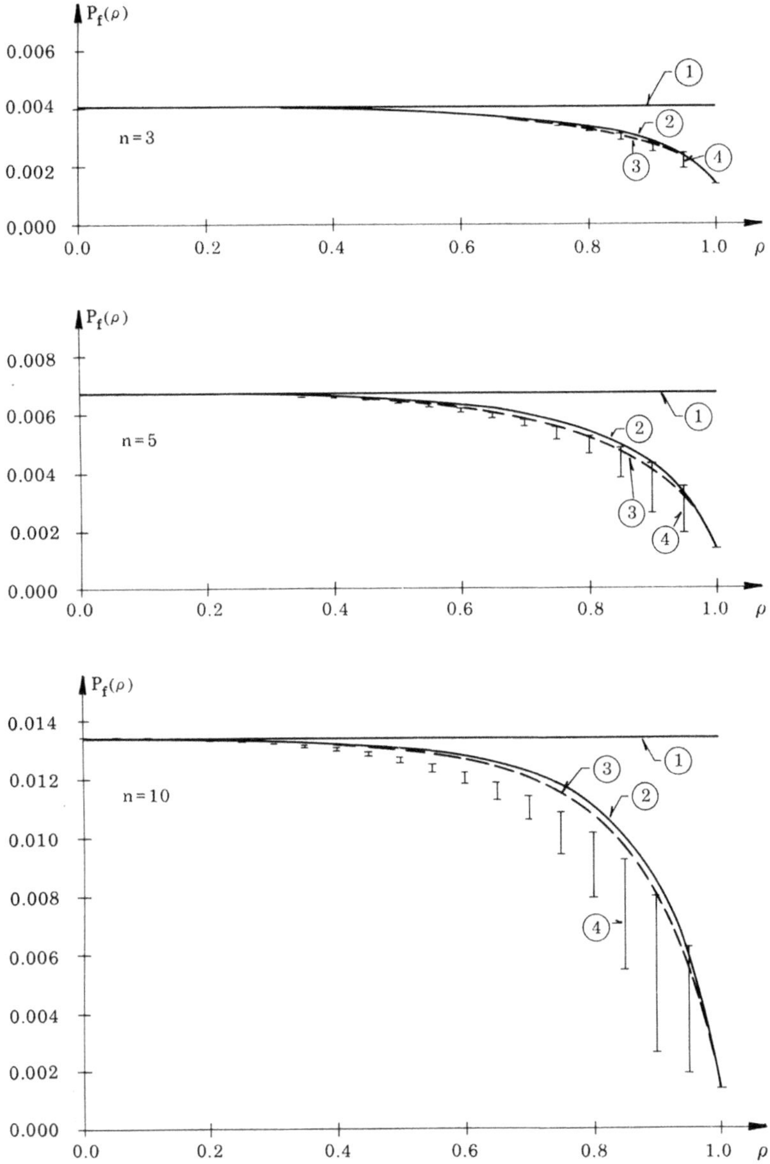

Figure 8.5

$$\Phi(-\bar{\bar{\beta}}) = 1-(1-\Phi(-\beta_1))(1-\Phi(-\beta_2)) \ldots (1-\Phi(-\beta_n)) \tag{8.33}$$

where Φ is the distribution function for the standard normal distribution and β_i the reliability index for element i. Note that $\bar{\bar{\beta}}$ corresponds to assuming that the parallel systems are mutually independent (see (7.3)). Therefore, using $\bar{\bar{\beta}}$ is on the safe side.

Example 8.6. Consider a series system with 5 elements satisfying the assumptions (1) and (2) on page 136, but omitting assumptions (3) and (4). Let the reliability indices for the five elements be

$$\beta_1 = 2.50,\ \beta_2 = 3.00,\ \beta_3 = 3.20,\ \beta_4 = 3.40,\ \beta_5 = 3.50$$

From (8.33) the equivalent reliability index $\bar{\bar{\beta}}$ can be calculated using a table for Φ. One gets

$$\Phi(-\bar{\bar{\beta}}) = 1-(1-62.10\cdot10^{-4})\,(1-13.50\cdot10^{-4})(1-6.87\cdot10^{-4})(1-3.37\cdot10^{-4})$$

$$(1-2.33\cdot10^{-4}) = 0.991202$$

and

$$\bar{\bar{\beta}} = -\,\Phi^{-1}(0.991202) = 2.37$$

Further, assume that the correlation matrix is the same as in example 8.4. The average correlation coefficient is then $\bar{\rho} = 0.28$ and by numerical integration an approximate value P_f' for the probability of failure can be calculated.

$$P_f' = 1-\int_{-\infty}^{\infty}[\Phi(\frac{2.37+\sqrt{0.28}\,t}{\sqrt{1-0.28}})]^5\,\varphi(t)dt \approx 43\cdot10^{-3}$$

Example 8.7. Consider a series system of two parallel systems with n_1 and n_2 elements, respectively, as shown in figure 8.6. Further assume that the strengths of the elements are identically distributed with the common mean μ and standard deviation σ and that the correlation coefficient between any pair of elements is ρ. Finally, assume that m elements are common to the two parallel systems. In this case the coefficient of correlation ρ_S between the two parallel systems can be calculated in the following way. Let R_1 and R_2 be the strengths of the two parallel systems. Then $\mu_{R_1} = n_1\mu$, $\mu_{R_2} = n_2\mu$ and

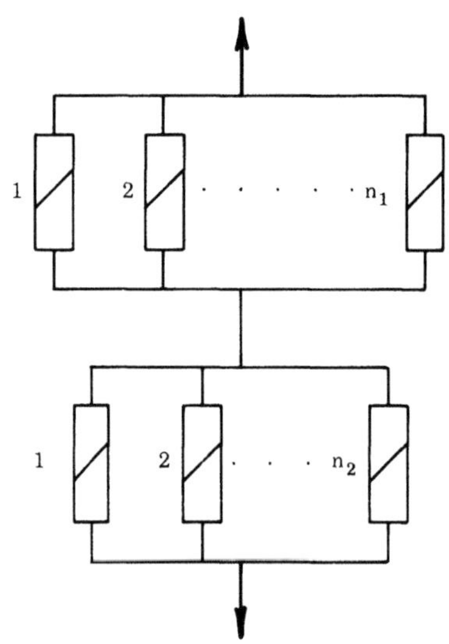

Figure 8.6

$$\mathrm{Var}[R_1] = n_1\sigma^2 + n_1(n_1 - 1)\rho\sigma^2$$

$$\mathrm{Var}[R_2] = n_2\sigma^2 + n_2(n_2 - 1)\rho\sigma^2$$

$$\mathrm{Cov}[R_1, R_2] = m\sigma^2 + (n_1 n_2 - m)\rho\sigma^2$$

Therefore

$$\rho_S = \frac{m + (n_1 n_2 - m)\rho}{\sqrt{(n_1 + n_1(n_1 - 1)\rho)(n_2 + n_2(n_2 - 1)\rho)}} \tag{8.34}$$

Exercise 8.2. Determine the coefficient of correlation ρ_S between two parallel systems with $n_1 = n_2 = 4$ elements and $m = 2$ common elements. Assume that the elements have identically distributed strength and common correlation coefficient $\rho = 0.8$.

BIBLIOGRAPHY

[8.1] Cornell, C. A.: *Bounds on the Reliability of Structural Systems.* J. Struct. Div., ASCE, Vol. 93, 1967, pp. 171 - 200.

[8.2] Ditlevsen, O.: *Narrow Reliability Bounds for Structural Systems.* J. Struct. Mech., Vol. 7, 1979, pp. 435 - 451.

[8.3] Hohenbichler, M. & R. Rackwitz: *Non-normal Dependent Vectors in Structural Safety.* J. Eng. Mech., ASCE, Vol. 107, 1981.

[8.4] Rackwitz, R.: *Close Bounds for the Reliability of Structural Systems.* Berichte zur Zuverlässigkeitstheorie der Bauwerke, SFB 96, Heft 29/1978, LKI, Technische Universität München.

[8.5] Thoft-Christensen, P. & J. D. Sørensen: *Reliability of Structural Systems with Correlated elements.* Applied Mathematical Modelling, Vol. 6, 1982.

Chapter 9

INTRODUCTION TO STOCHASTIC PROCESS THEORY AND ITS USES

9.1 INTRODUCTION

In the preceding chapters, loads and strengths have mainly been modelled by random variables with associated distribution functions. However, a load S on a given structure will usually be time-varying S(t). The function S(t) is stochastic (random) in the sense that the value of S at a given time t is an outcome of a random variable. In this way, by modelling the time history and the randomness of a physical quantity by an (infinite) number of random variables, a so-called *stochastic process* is obtained. In section 9.2 a more formal definition of this concept will be given, but it is not possible to give a detailed treatment of the theory of stochastic processes here. Only the most fundamental notions will be introduced and only one special type of stochastic processes will be described in more detail.

A very important problem in relation to a stochastic process is the *barrier crossing* problem. Consider, for example, the response of a structure expressed by the time-history of a given stress. When modelling the time-history of the stress by a stochastic process it might be of interest to evaluate the probability that the process stays within specified bounds during the expected lifetime of the structure. This problem will also be briefly examined.

9.2 STOCHASTIC PROCESSES

As mentioned above a *stochastic process* is an indexed set $\{X(t), t \in T\}$ of random variables X(t), where all X(t) are defined on the same sample space Ω. Note that two different kinds of variables are involved, namely the stochastic variables X(t) and the variable t, here called the index. The *index set T* is typically a time-interval, but can be any kind of finite set, a countably infinite set or a subset of R. For the sake of simplicity t will be assumed in the following to be the variable time.

The probabilistic structure of a stochastic process can be described in a way similar to random vectors. If the index set is a finite set then the stochastic process forms a random vector. The fact that a stochastic process is a set of random variables makes it natural to describe its probabilistic structure in a way similar to random vectors, but in this case the index set is infinite.

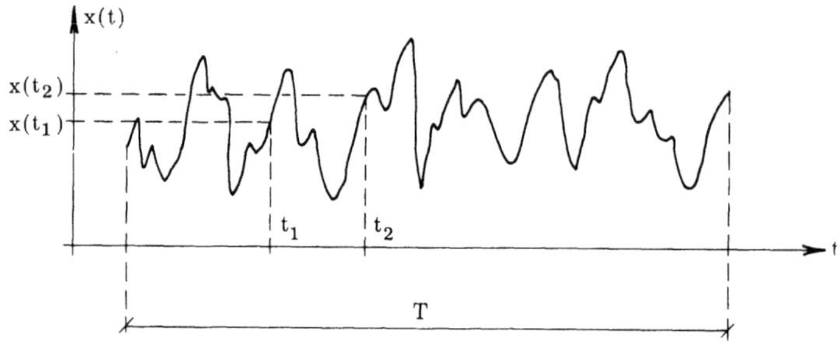

Figure 9.1

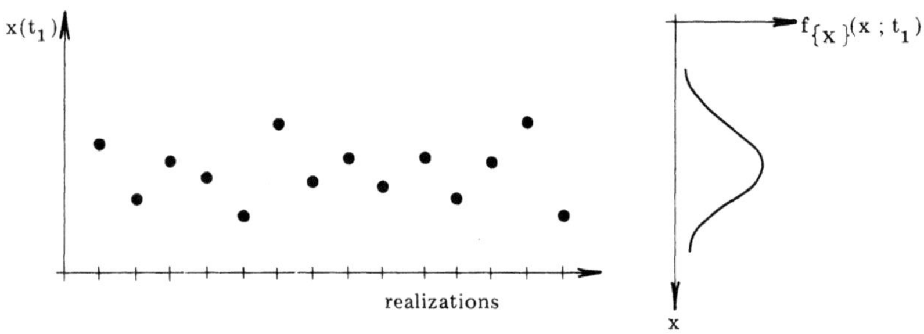

Figure 9.2

For a fixed sample space Ω the outcome of the set of stochastic variables form an ordinary function called a *realization* (see figure 9.1). A realization might be thought of as the outcome of an experiment. If the experiment is repeated, the new realisations will not be the same as in figure 9.1, but its probabilistic contents will be the same. In figure 9.2 values of $X(t_1)$ are shown for a number of realizations and also the associated density function $f_{\{X\}}(x\,;t_1)$.

Given two instants of time t_1 and t_2 some correlation between $x(t_1)$ and $x(t_2)$ will usually exist, especially when the time-difference $|t_1 - t_2|$ is small. This is taken into consideration through the joint distribution function $F_{\{X\}}(x_1, x_2\,; t_1, t_2)$ defined (see (2.58)) by

$$F_{\{X\}}(x_1, x_2\,; t_1, t_2) = P((X(t_1) \leqslant x_1) \cap (X(t_2) \leqslant x_2)) \tag{9.1}$$

This joint distribution function for arbitrary $(t_1, t_2) \in T^2$ is called the *joint distribution function of order 2.* The corresponding *joint density function of order 2* is given by

$$f_{\{X\}}(x_1, x_2\,; t_1, t_2) = \frac{\partial^2 F_{\{X\}}(x_1, x_2\,; t_1, t_2)}{\partial x_1 \partial x_2} \tag{9.2}$$

The definitions (9.1) and (9.2) can easily be generalized to probability functions of any order n, $n = 3, 4, \ldots$

In describing a stochastic process the following functions (of time) are of great interest. The *mean value function* $\mu_X(t)$ is defined as the expected value of X(t)

$$\mu_X(t) = E[X(t)] = \int_{-\infty}^{\infty} x f_{\{X\}}(x; t) dx \tag{9.3}$$

The *autocorrelation function* $R_{XX}(t_1, t_2)$ is equal to the following joint moment of the random variables $X(t_1)$ and $X(t_2)$

$$R_{XX}(t_1, t_2) = E[X(t_1)X(t_2)] = \int_{-\infty}^{\infty}\int_{-\infty}^{\infty} x_1 x_2 f_{\{X\}}(x_1, x_2; t_1, t_2) dx_1 dx_2 \tag{9.4}$$

The *autocovariance function* $C_{XX}(t_1, t_2)$ is the covariance of the random variables $X(t_1)$ and $X(t_2)$

$$C_{XX}(t_1, t_2) = E[(X(t_1) - \mu_X(t_1))(X(t_2) - \mu_X(t_2))]$$
$$= R_{XX}(t_1, t_2) - \mu_X(t_1)\mu_X(t_2) \tag{9.5}$$

By setting $t_1 = t_2 = t$ in (9.5), the *variance function* $\sigma_X^2(t)$ of the random variables X(t) is obtained

$$\sigma_X^2(t) = C_{XX}(t, t) = R_{XX}(t, t) - \mu_X^2(t) \tag{9.6}$$

Finally the *autocorrelation coefficient* $\rho_{XX}(t_1, t_2)$ is defined in a similar manner to (2.80) by

$$\rho_{XX}(t_1, t_2) = \frac{C_{XX}(t_1, t_2)}{\sigma_X(t_1)\sigma_X(t_2)} \tag{9.7}$$

For an important group of stochastic processes all finite dimensional distributions are invariant to a linear translation of the index set. This can also be expressed by the statement that all distributions are invariant to a translation of the time origin. Such processes are called *strictly homogeneous* or when the index parameter is time, *strictly stationary*.

When this invariant assumption only holds for distributions of order one and two the process is called *weakly homogeneous* or *weakly stationary*. In the following, the word stationary will be used in the last-mentioned meaning.

An important consequence of the assumption of stationarity is that $f_{\{X\}}(x; t)$ and $F_{\{X\}}(x; t)$ becomes independent of t so that we can omit reference to t. Further, the second-order distributions (9.1) will only depend on the difference of the index parameter $\tau = t_1 - t_2$. The same is true for all the other statistics mentioned above.

In practical applications, the modelling of a physical quantity by a stochastic process must often be based on a single realisation of a stationary process. If only one realisation is at hand it is natural to estimate the mean value in the following way

$$\mu = \frac{1}{T}\int_0^T x(\tau)d\tau \tag{9.8}$$

If this time average approaches μ_X for $T \to \infty$ the process is said to be *ergodic in the mean value*. In the same manner a process is *ergodic in correlation* if

$$R(\tau) = \frac{1}{T-\tau}\int_0^{T-\tau} x(t+\tau)x(t)dt \tag{9.9}$$

approaches $R_{XX}(\tau)$ for $T \to \infty$. If this property holds for all moments, the process is called *ergodic*.

Note that stationarity is an assumption behind the definition of an ergodic process so that any ergodic process is stationary but not vice versa.

9.3 GAUSSIAN PROCESSES

In this section so-called Gaussian processes are treated. It has been stated several times that a linear transformation of a set of Gaussian (normal) random variables result in a new set of Gaussian random variables. This important property of Gaussian random variables is the main reason why they are used for modelling whenever it can be justified. In a similar manner, it can be shown that linear operations on a Gaussian process results in another Gaussian process.

A process $\{X(t), t \in T\}$ is *Gaussian* if the random variables $X(t_1), X(t_2), \ldots, X(t_n)$ are jointly normal for any $n, t_1, t_2, \ldots, t_n$. The probability density function for the corresponding n-dimensional n^{th} order distribution is then given by (see (2.89))

$$f_{\{X\}}(x_1, \ldots, x_n; t_1, \ldots, t_n) =$$

$$\frac{1}{(2\pi)^{\frac{n}{2}} |\bar{\bar{C}}|^{\frac{1}{2}}} e^{-\frac{1}{2}\sum_{i,j=1}^{n}(x_i - \mu_X(t_i))M_{ij}(x_j - \mu_X(t_j))} \tag{9.10}$$

where $\bar{\bar{C}}$ is the autocovariance matrix

$$\bar{\bar{C}} = \begin{bmatrix} C_{XX}(t_1,t_1) & C_{XX}(t_1,t_2) & \cdots & C_{XX}(t_1,t_n) \\ \vdots & \vdots & & \vdots \\ C_{XX}(t_n,t_1) & C_{XX}(t_n,t_2) & \cdots & C_{XX}(t_n,t_n) \end{bmatrix} \tag{9.11}$$

and M_{ij} is the i, jth element in $\bar{\bar{C}}^{-1}$. It is clear from the definition (9.10) that a Gaussian process is completely determined by the mean value function $\mu_X(t)$ and the autocovariance function $C_{XX}(t_1, t_2)$. Therefore, a stationary Gaussian process is always strictly stationary. An important property of a Gaussian process $\{X(t)\}$ is that its derivative process $\{\dot{X}(t)\}$ is also a Gaussian process. Let x(t) be a realization of $\{X(t)\}$ and let

$$\dot{x}(t) = \frac{d}{dt} x(t) \tag{9.12}$$

be meaningful. The derivative process $\{\dot{X}(t)\}$ is then determined by the realizations $\dot{x}(t)$ when almost all realizations x(t) of $\{X(t)\}$ are considered.

Example 9.1. Consider two independent normal random variables X_1 and X_2 with $\mu_{X_1} = \mu_{X_2} = 0$ and $\sigma^2_{X_1} = \sigma^2_{X_2} = \sigma^2$. Let a stochastic process $\{X(t)\}$ be given by

$$X(t) = X_1 \cos(\omega t) + X_2 \sin(\omega t) \tag{9.13}$$

where ω is a constant. The random variables $X(t_i)$, $t_i \in T$ are clearly jointly normal and their statistics are determined by the mean and autocorrelation of the process $\{X(t)\}$. By (9.13)

$$\mu_X(t) = E[X(t)] = 0 \tag{9.14}$$

and by the definition (9.4),

$$R_{X_1X_2}(t_1, t_2) = E[(X_1 \cos\omega t_1 + X_2 \sin\omega t_1)(X_1 \cos\omega t_2 + X_2 \sin\omega t_2)]$$

$$= E[X_1^2]\cos\omega t_1 \cos\omega t_2 + E[X_2^2]\sin\omega t_1 \sin\omega t_2 = \sigma^2 \cos\omega(t_1 - t_2) \tag{9.15}$$

since $E[X_1X_2] = 0$. From (9.15)

$$\sigma_X^2(t) = R_{XX}(t, t) - \mu_X^2(t) = \sigma^2 \tag{9.16}$$

The process $\{X(t)\}$ is therefore a stationary Gaussian process with zero mean and variance σ^2.

Example 9.2. Consider the same process $\{X(t)\}$ as in example 9.1. The autocorrelation coefficient is

$$\rho_{XX}(t_1, t_2) = \frac{R_{XX}(t_1, t_2)}{\sigma_X(t_1)\sigma_X(t_2)} = \cos\omega(t_1 - t_2) \tag{9.17}$$

so that the joint distribution density function is given by

$$f_{\{X\}}(x_1, x_2 ; t_1, t_2) = \frac{1}{2\pi\sigma^2 \sqrt{1 - \cos^2 \omega\tau}} e^{-\frac{x_1^2 - 2x_1x_2\cos\omega\tau + x_2^2}{2\sigma^2(1 - \cos^2\omega\tau)}} \tag{9.18}$$

where $\tau = \tau_1 - \tau_2$.

An important property of the autocorrelation function $R_{XX}(\tau)$ of a stationary stochastic process $\{X(t)\}$ is the following. If $R_{XX}(\tau)$ has a second derivative $R''_{XX}(\tau)$ which is continuous at $\tau = 0$ then the derivative process $\{\dot{X}(t)\}$, defined by its realizations by (9.12), is also a stationary stochastic process. And it can be shown that

$$E[\dot{X}^2] = -R''_{XX}(0) \tag{9.19}$$

and

$$E[X\dot{X}] = 0 \tag{9.20}$$

so that there is no correlation between $\{X(t)\}$ and $\{\dot{X}(t)\}$. Further $E[\dot{X}] = \frac{d}{dt} E[X]$.

Example 9.3. Let $\{X(t)\}$ be a stationary Gaussian process with zero mean. It follows then from the remarks above that the joint density function $f_{\{X\}\{\dot{X}\}}$ is

$$f_{\{X\}\{\dot{X}\}}(x, \dot{x}) = \frac{1}{2\pi\sigma_X\sigma_{\dot{X}}} e^{-\frac{x^2}{2\sigma_X^2} - \frac{\dot{x}^2}{2\sigma_{\dot{X}}^2}} \tag{9.21}$$

where $\sigma_{\dot{X}}^2 = E[\dot{X}^2] - E[\dot{X}]^2 = E[\dot{X}^2] = -R''_{XX}(0)$.

9.4 BARRIER CROSSING PROBLEM

In this section it will be shown for a stochastic process $\{X(t)\}$ how the number of crossings of a given barrier (threshold) in a given time-interval can be estimated. The presentation here is in accordance with the book by Lin. Figure 9.3 shows a realization $x(t)$ in the interval $[t_1 \,;\, t_2]$ of a stochastic process $\{X(t)\}$ and a constant barrier $x(t) = \xi$. The number of upcrossings of this barrier in the time interval $[t_1 \,;\, t_2]$ is four. In the following an upcrossing will be called a *positive passage* and a downcrossing a *negative passage*.

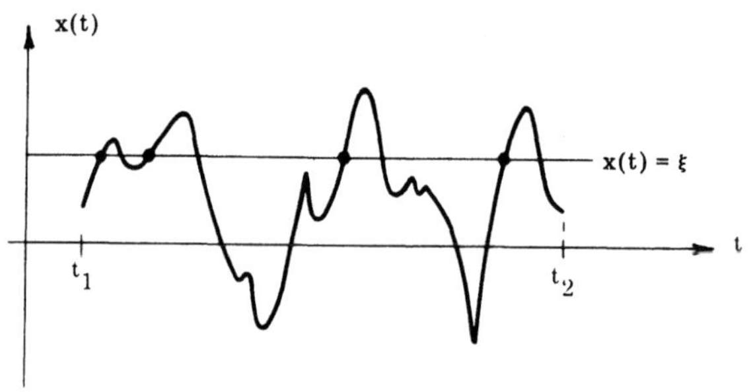

Figure 9.3

To solve the problem of estimating the expected number of positive passages of a given barrier the so-called *Heaviside step function* H is a useful too. Heaviside's step function H is defined by (see figure 9.4)

$$H(x) = \begin{cases} 0 & \text{for} \quad x < 0 \\ \frac{1}{2} & \text{for} \quad x = 0 \\ 1 & \text{for} \quad x > 0 \end{cases} \tag{9.22}$$

By formal differentiation of the function H one gets the so-called *Dirac delta function* $\delta(x)$. $\delta(x)$ is not an ordinary function in the sense that a definite value can be assigned to every x. For our purposes, it can be defined by

$$\delta(x) = \lim_{\epsilon \to 0} \frac{1}{\sqrt{2\pi}\,\epsilon} e^{-\frac{x^2}{2\epsilon^2}} \tag{9.23}$$

What is required here, is only the property that integration of $\delta(x)$ gives H(x).

For a stochastic process $\{X(t)\}$ and a given barrier $x(t) = \xi$, it is then convenient to define a new stochastic process $\{Y(t)\}$ by

$$Y(t) = H(X(t) - \xi) \tag{9.24}$$

or

$$y(t) = \begin{cases} 0 & \text{for} \quad x(t) < \xi \\ \frac{1}{2} & \text{for} \quad x(t) = \xi \\ 1 & \text{for} \quad x(t) > \xi \end{cases} \tag{9.25}$$

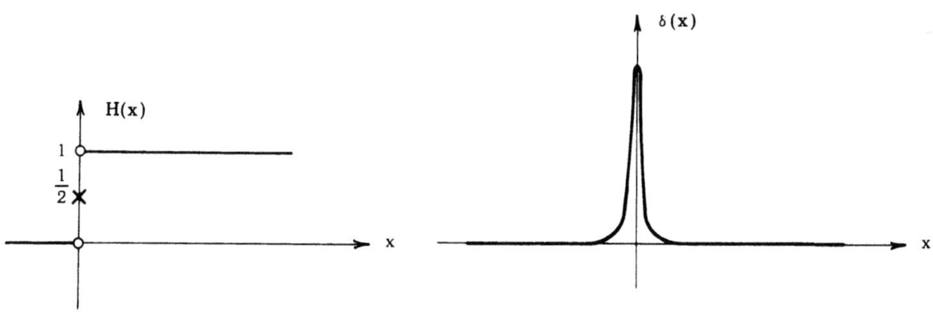

Figure 9.4

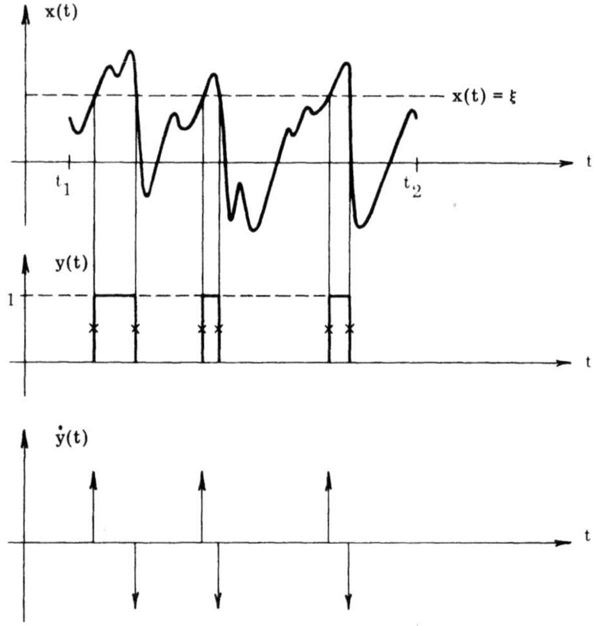

Figure 9.5

By formal differentiation of (9.24) the derivative process $\{\dot{Y}\}$ can be determined by

$$\dot{Y}(t) = \dot{X}(t)\delta(X(t) - \xi) \tag{9.26}$$

where the existence of $\dot{X}(t)$ is assumed. For a realization x(t) of the process $\{X(t)\}$ the corresponding realizations y(t) and $\dot{y}(t)$ of the processes $\{Y(t)\}$ and $\{\dot{Y}(t)\}$ are shown in figure 9.5.

Note that the realization $\dot{y}(t)$ consists of a number of unit impulses. A positive unit impulse corresponds to a positive passage of the barrier and a negative unit impulse corresponds to a negative passage of the barrier. These impulses are unit impulses because integration of $\dot{y}(t)$ over one impulse must yield +1 or −1.

By counting the number of such unit impulses in the time interval $[t_1 ; t_2]$ the total number n of crossings of the barrier $x(t) = \xi$ is obtained. This can also be formulated in the following way

$$n(\xi, t_1, t_2) = \int_{t_1}^{t_2} |\dot{y}(t)| dt = \int_{t_1}^{t_2} |\dot{x}(t)| \cdot \delta(x(t) - \xi) dt \tag{9.27}$$

From (9.27), the number $n(\xi, t_1, t_2)$ of crossings or passages of a given barrier can be calculated for any realization $x(t)$ of the stochastic process $\{X(t)\}$. Such a set of numbers can be considered the outcome of a random variable $N(\xi, t_1, t_2)$. The expected number of crossings can now be determined.

$$E[N(\xi, t_1, t_2)] = \int_{t_1}^{t_2} E[\,|\dot{X}(t)| \cdot \delta(X(t) - \xi)]dt =$$

$$\int_{t_1}^{t_2} \int_{-\infty}^{\infty} \int_{-\infty}^{\infty} |\dot{x}| \delta(x - \xi) f_{X\dot{X}}(x, \dot{x}; t) dx d\dot{x} dt =$$

$$\int_{t_1}^{t_2} \int_{-\infty}^{\infty} |\dot{x}| f_{X\dot{X}}(\xi, \dot{x}; t) d\dot{x} dt \tag{9.28}$$

where $f_{X\dot{X}} = f_{\{X\}\{\dot{X}\}}$ is the joint density function for $\{X(t)\}$ and $\{\dot{X}(t)\}$.

It is convenient to consider the *rate of crossings per unit time* N' instead of the number of crossings N in the time interval considered. N' and N are related in the following way

$$N(\xi, t_1, t_2) = \int_{t_1}^{t_2} N'(\xi, t) dt \tag{9.29}$$

Equation (9.28) can then be written in the more simple form

$$E[N'(\xi, t)] = \int_{-\infty}^{\infty} |\dot{x}| f_{X\dot{X}}(\xi, \dot{x}; t) d\dot{x} \tag{9.30}$$

Now assume that the stochastic process $\{X(t)\}$ is *stationary* so that $f_{X\dot{X}}$ is independent of the time t. Then

$$E[N'(\xi, t)] = \int_{-\infty}^{\infty} |\dot{x}| f_{X\dot{X}}(\xi, \dot{x}) d\dot{x} = K(\xi) \tag{9.31}$$

so that the expected rate of crossing per unit time $E[N'(\xi, t)]$ is independent of time, but of course dependent on the barrier ξ. The expected total number of crossings in the time interval $[t_1 ; t_2]$ is therefore (see (9.29))

$$E[N(\xi, t_1, t_2)] = K(\xi)(t_2 - t_1) \tag{9.32}$$

Using equations (9.31) and (9.32), the number of crossings of the barrier ξ is determined, i.e. upcrossings (positive crossings) as well as downcrossings (negative crossings). But for a stationary stochastic process it is reasonable to assume that any positive crossing is followed by a negative crossing. Therefore

$$E[N'_+(\xi)] = E[N'_-(\xi)] = \frac{1}{2} E[N'(\xi)] \tag{9.33}$$

where $N'_+(\xi)$ is the rate of positive crossings of the barrier ξ and $N'_-(\xi)$ is the rate of negative crossings of the barrier ξ.

Note that for positive crossings $\dot{x} > 0$ so that from (9.30)

$$E[N'_+(\xi)] = \int_0^\infty \dot{x}\, f_{X\dot{X}}(\xi, \dot{x})\, d\dot{x} \tag{9.34}$$

and similarly for negative crossings. Hence, for stationary processes, the joint density function $f_{X\dot{X}}$ is an even function in the variable $\dot{x}$. The fundamental formula (9.34) is called *Rice's formula.*

Example 9.4. Let $\{X(t)\}$ be a stationary Gaussian process with zero mean. The joint density function $f_{X\dot{X}}$ is then given by (9.21). From Rice's formula (9.34)

$$E[N'_+(\xi)] = \int_0^\infty \dot{x}\, \frac{1}{2\pi\sigma_X\sigma_{\dot{X}}}\, e^{-\frac{\xi^2}{2\sigma_X^2} - \frac{\dot{x}^2}{2\sigma_{\dot{X}}^2}}\, d\dot{x} = \frac{1}{2\pi}\frac{\sigma_{\dot{X}}}{\sigma_X}\, e^{-\frac{\xi^2}{2\sigma_X^2}} \tag{9.35}$$

For $\xi = 0$ the expected rate of positive zero crossings is

$$E[N'_+(0)] = \frac{1}{2\pi}\frac{\sigma_{\dot{X}}}{\sigma_X} \tag{9.36}$$

Example 9.5. Consider a stationary non-Gaussian process $\{X(t)\}$ with the following joint density function

$$f_{X\dot{X}}(x, \dot{x}) = \begin{cases} \frac{1}{2}(1 + x) & \text{for} \quad (x, \dot{x}) \in [-1; 0] \times [-1; 1] \\ \frac{1}{2}(1 - x) & \text{for} \quad (x, \dot{x}) \in [0; 1] \times [-1; 1] \\ 0 & \text{otherwise} \end{cases} \tag{9.37}$$

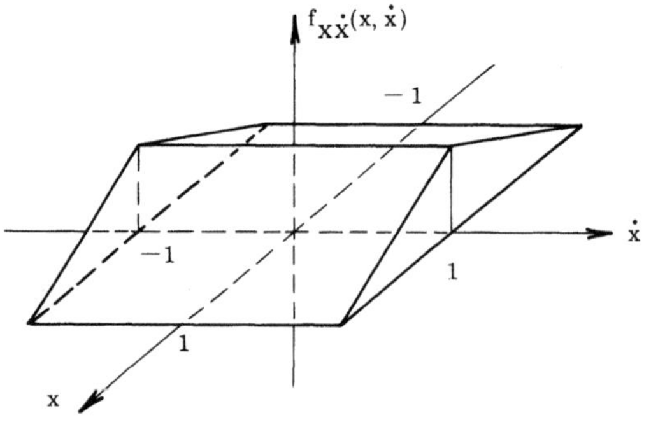

Figure 9.6

The expected rate of positive crossings of the barrier x(t) = ξ is given by (9.34)

$$E[N'_+(\xi)] = \int_0^1 \dot{x}\,\frac{1}{2}(1 \pm \xi)d\dot{x} = \begin{cases} \frac{1}{4}(1+\xi) & \text{for } -1 \leqslant \xi \leqslant 0 \\ \frac{1}{4}(1-\xi) & \text{for } 0 \leqslant \xi \leqslant 1 \\ 0 & \text{otherwise} \end{cases} \tag{9.38}$$

Example 9.6. Consider the same stochastic process {X(t)} as in example 9.5, but in this case the joint density function (9.37) is approximated by a 2-dimensional normal distribution in such a way that the two marginal density functions have the same means and variances.

The marginal density functions for the distribution (9.37) are shown in figure 9.7. It is then easy to see that

$$\begin{aligned} &\mu_X = \mu_{\dot{X}} = 0 \\ &\sigma_X^2 = \frac{1}{6} \quad , \quad \sigma_{\dot{X}}^2 = \frac{1}{3} \end{aligned} \tag{9.39}$$

The approximate normal distribution is therefore

$$f_{X\dot{X}}(x, \dot{x}) = \frac{3\sqrt{2}}{2\pi}\, e^{-3x^2 - 1.5\dot{x}^2} \tag{9.40}$$

The expected rate of positive zero crossings for the corresponding stationary Gaussian process is

$$E[N'_+(0)] = \frac{1}{2\pi}\frac{\sqrt{6}}{\sqrt{3}} = 0.2251 \tag{9.41}$$

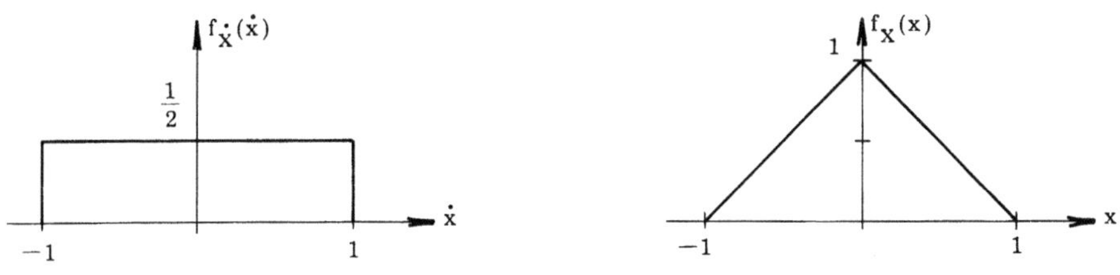

Figure 9.7

Exercise 9.1. Approximate the joint density function (9.37) by a 2-dimensional normal distribution in such a way that the rate of positive crossings of the barriers ξ = 0 and ξ = 1/2 is equal for the corresponding stochastic processes.

(Answer $f_{X\dot{X}}(x, \dot{x}) = 0.56\, e^{-2.77x^2 - 1.12\dot{x}^2}$).

9.5 PEAK DISTRIBUTION

The results derived in section 9.4 can be used to investigate the statistics of the peak distribution of a stochastic process $\{X(t)\}$, because peaks or troughs (extrema in $\{X(t)\}$) occur when the stochastic process $\{\dot{X}(t)\}$ has a zero crossing. The number of zero crossings of $\{\dot{X}(t)\}$ is equal to the number of extrema in $\{X(t)\}$. The formulas derived in section 9.4 can therefore be used when $\{\dot{X}(t)\}$ and $\{\ddot{X}(t)\}$ are substituted for $\{X(t)\}$and $\{\dot{X}(t)\}$.

When the process $\{X(t)\}$ is a *narrow-band Gaussian process* the distribution of the peaks can be determined in a very simple way. A realization of a narrow-band process is shown in figure 9.8. It is similar to a sinusoid, but the amplitude and phase are slowly varying. The stationary response of a lightly damped linear system will often be narrow-banded, when the input process is a broad-banded Gaussian process, such as an earthquake excitation.

In this case the expected number of peaks above the level $\xi(\xi > 0)$ per unit time is, with good approximation, equal to the expected rate of crossings of the barrier ξ, i.e. equal to $E[N'_+(\xi)]$. Similarly, the expected total number of peaks per unit time is equal to the expected rate of zero crossings $E[N'_+(0)]$. Therefore, the expected relative number of peaks above ξ per unit time is

$$\frac{E[N'_+(\xi)]}{E[N'_+(0)]} = e^{-\frac{\xi^2}{2\sigma_X^2}} \tag{9.42}$$

where the formulas (9.35) and (9.36) have been used.

The distribution function $F_\Xi(\xi)$ for the peak magnitude $(\xi > 0)$ is then given by

$$F_\Xi(\xi) = 1 - e^{-\frac{\xi^2}{2\sigma_X^2}} \quad , \quad 0 \leqslant \xi < \infty \tag{9.43}$$

and the density function $f_\Xi(\xi)$ by

$$f_\Xi(\xi) = \frac{\xi}{\sigma_X^2} e^{-\frac{\xi^2}{2\sigma_X^2}} \quad , \quad 0 \leqslant \xi < \infty \tag{9.44}$$

This distribution is the so-called Rayleigh distribution.

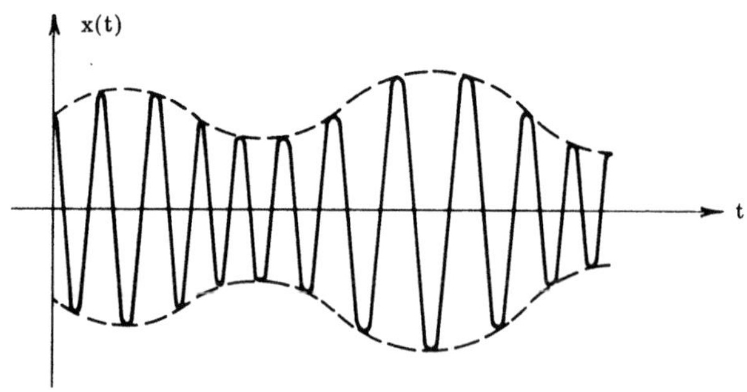

Figure 9.8

Example 9.7. Let $\{X(t)\}$ be a narrow-band Gaussian process with zero mean and let $\sigma_X = 1$. The density function for the peak magnitude is then given by (9.44)

$$f_{\Xi}(\xi) = \xi \, e^{-\frac{1}{2}\xi^2} \tag{9.45}$$

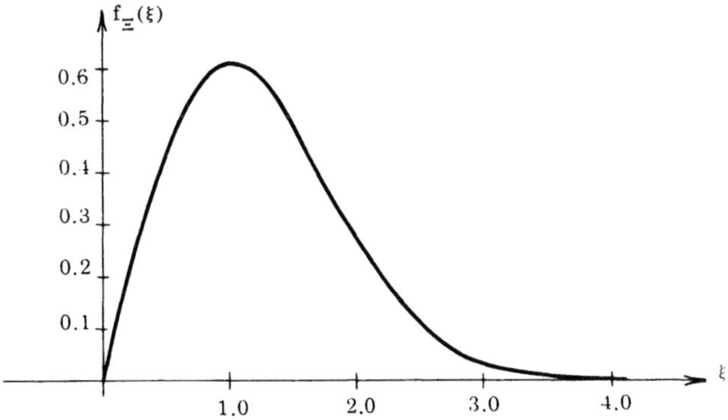

Figure 9.9. Density function (9.45).

Example 9.8. From an experimental investigation of the variation in bending moment with time in a given section of a beam it is concluded that the moment can be modelled by a stationary narrow-band Gaussian process $\{M(t)\}$ with

$$\mu_M = 12 \text{ MNm} \quad ; \quad \sigma_M = 2 \text{ MNm}$$

$$\mu_{\dot{M}} = 0 \text{ MNm/sec} \quad ; \quad \sigma_{\dot{M}} = 4.01 \cdot 10^{-2} \text{ MNm/sec}$$

The rate of positive crossings of the barrier $\xi = 18$ MNm is then

$$E[N'_+(18)] = \frac{1}{2\pi} \frac{4.01 \cdot 10^{-2}}{2} e^{-\frac{(18-12)^2}{2 \cdot 4}} = 3.54 \cdot 10^{-5} \text{ sec}^{-1} \tag{9.46}$$

The density function for the peak magnitude is given by

$$f_{\Xi}(\xi) = \frac{1}{4}(\xi - 12) e^{-\frac{1}{2 \cdot 4}(\xi - 12)^2} \tag{9.47}$$

and the probability of getting peak magnitudes greater than $\xi = 18$ MNm is

$$P(\xi > 18) = 1 - F_{\Xi}(18) = e^{-\frac{1}{2 \cdot 4}(18-12)^2} = 0.011 \tag{9.48}$$

In the derivations above, only narrow-banded processes are considered, i.e. processes where the ratio

$$\alpha = \frac{\text{expected number of zero crossings}}{\text{expected number of peaks}} \tag{9.49}$$

is approximately equal to 1 (see figure 9.8). It can be shown that the ratio α lies between zero and one and that in general the density function $f_{\Xi}(\xi)$ for the peak magnitude is given by

$$f_{\Xi}(\xi) = \frac{1}{\sqrt{2\pi}} \frac{\sqrt{1-\alpha^2}}{\sigma_X} e^{-\frac{\xi^2}{2\sigma_X^2(1-\alpha^2)}} + \frac{\alpha\xi}{2\sigma_X^2}[1 + \mathrm{erf}(\frac{\xi}{\sigma_X}(\frac{2}{\alpha^2}-2)^{-\frac{1}{2}})]e^{-\frac{\xi^2}{2\sigma_X^2}} \tag{9.50}$$

where the error function erf is defined by

$$\mathrm{erf}(x) = \frac{2}{\sqrt{\pi}}\int_0^x e^{-t^2}\,dt \tag{9.51}$$

Exercise 9.2. Show that (9.50) for $\alpha = 1$ is equal to (9.44).

When α is very small ($\alpha \approx 0$) corresponding to a large number of peaks in relation to the number of zero crossings the formula (9.50) can be approximated by a normal distribution

$$f_{\Xi}(\xi) \approx \frac{1}{\sqrt{2\pi}\,\sigma_X} e^{-\frac{\xi^2}{2\sigma_X^2}}, \quad -\infty < \xi < \infty \tag{9.52}$$

Example 9.8. Consider an ergodic Gaussian process $\{X(t)\}$ with $\mu_X = 5$ and $\sigma_X = 2$. By analysing a realization of this process it is concluded that the ratio α of the expected number of zero crossings to expected number of peaks can be set equal to 0.6. Further, the expected number of crossings of the barrier $\xi = 9$ is equal to 10^{-3}.

The probability of getting peak magnitudes greater than 9 is then

$$P(\xi > 9) = 1 - \int_{-\infty}^{9} f_{\Xi}(\xi)d\xi \tag{9.53}$$

where

$$f_{\Xi}(\xi) = 0.160\cdot e^{-0.195(\xi-5)^2} + 0.075(\xi-5)[1 + \mathrm{erf}(0.265\xi)]e^{-\frac{(\xi-5)^2}{8}} \tag{9.54}$$

By numerical integration $P(\xi > 9)$ can then be calculated from (9.53). Lower and upper bounds for $P(\xi > 9)$ can be calculated by considering the cases $\alpha = 0$ and $\alpha = 1$.

For $\alpha = 0$, one gets from (9.52)

$$P(\xi > 9) = 1 - \Phi(\frac{9-5}{2}) = 0.02275$$

and for $\alpha = 0$ from (9.44)

$$P(\xi > 9) = e^{-\frac{1}{8}(9-5)^2} = 0.135$$

The standard deviation $\sigma_{\dot{X}}$ for the derivative process can be calculated by setting $E[N_+'(9)]$ equal to $\frac{1}{2}\cdot 10^{-3}$. One gets $\sigma_{\dot{X}} = 0.046$. The expected rate of positive crossings of any barrier can then be calculated from (9.35).

Exercise 9.3. Consider an ergodic narrow-banded Gaussian process $\{X(t)\}$. By analysing a realization of this process it is concluded that the expected rates of positive crossings of the barriers $\xi = 0$, 5, and 10 are 10^{-2}, 10^{-3}, and 10^{-5}, respectively. Determine the mean and the variance for $\{X(t)\}$ and for $\{\dot{X}(t)\}$. Sketch the density function for the peak magnitude

: and calculate the probability of obtaining peak values greater than 5.

: (Answer: $P(\xi > 5) = 0.075$).

BIBLIOGRAPHY

[9.1] Crandall, S. H. & W. D. Mark: *Random Vibration in Mechanical Systems.* Academic Press, N. Y., 1973.

[9.2] Krenk, S.: *First-Passage Times and Extremes of Stochastic Processes.* Lectures on Structural Reliability (ed. P. Thoft-Christensen), Aalborg University Centre, Aalborg, Denmark, 1980, pp. 75-98.

[9.3] Lin, Y. K.: *Probabilistic Theory of Structural Dynamics.* McGraw-Hill, N.Y. 1967.

[9.4] Nielsen, S. K.: *Strength- and Load Processes.* Lectures on Structural Reliability (ed. P. Thoft-Christensen), Aalborg University Centre, Aalborg, Denmark, 1980, pp. 39-73.

[9.5] Papoulis, A.: *Probability, Random Variables and Stochastic Processes.* McGraw-Hill, N. Y., 1965.

Chapter 10
LOAD COMBINATIONS

10.1 INTRODUCTION

The modelling of load variables is treated briefly in section 3.5. It is stressed there that load variables and other actions are typically time-varying quantities which are best modelled as stochastic processes. In section 3.5, it is also shown that when dealing with a *single time-varying load* in connection with barrier crossing problems (see section 9.4) the detailed time variation is not of relevance. This is due to the fact that in such cases the distribution of the maximum value of the loading process in a given reference periode can be derived from the arbitrary-point-in-time distribution (see figure 3.13 on page 57). When the loading process is continuous then the probability distribution of the maximum value (largest extreme) is likely to be very closely approximated by one of the asymptotic extreme value distributions, treated in section 3.3. In this way instead of modelling a single load variable as a stochastic process $\{X(t)\}$ it is modelled by a stochastic variable, say Y (see also section 9.5). Therefore, in reliability analysis, single load variables imply no special difficulties. A number of examples in chapters 5 and 6 of analysis and design of simple structures loaded by single loads illustrate this fact.

When more than one time-varying load variable acts in *combination* on a structure then the above simplification cannot be used because determination of the distribution of the combined load effect requires knowledge of the detailed variation with time of the individual loading processes. This is illustrated in figure 10.1, where realisations $p_1(t)$ and $p_2(t)$ of two loading processes $\{P_1(t), 0 \leq t \leq T\}$ and $\{P_2(t), 0 \leq t \leq T\}$ are shown together with the sum $p_1(t) + p_2(t)$. It is clear from figure 10.1 that the maximum values of $p_1(t)$, $p_2(t)$ and $p_1(t) + p_2(t)$ during the reference period need not appear at the same instant of time. For the specific realisation shown here, the instants of time t_1, t_2 and t_3 for maximum of $p_1(t)$, $p_2(t)$ and $p_1(t) + p_2(t)$ are all different. Also note that maximum value of $p_1(t) + p_2(t)$ is considerably smaller than the sum of the maximum values of $p_1(t)$ and $p_2(t)$. It is obvious from these observations that knowledge of the detailed time variation of the two loading variables in the reference period T is required to determine the probability distribution of the sum of the two load variables. Therefore, knowledge of the distribution of only the maximum values of the individual loading processes gives insufficient information to evaluate the combined effect exactly.

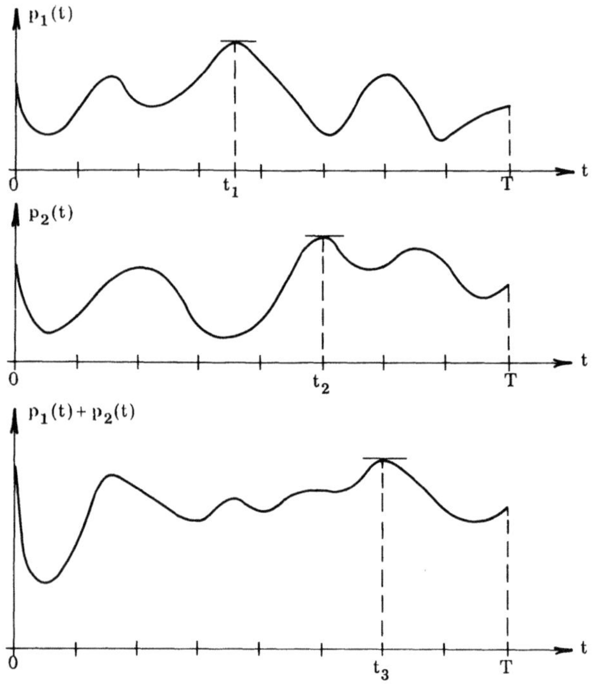

Figure 10.1

The intention of chapter 10 is to give some information on problems connected with load combinations. However, a thorough presentation of these problems is beyond the scope of this book. The reader is referred to the references at the end of the chapter. The main intention is to give the necessary background for understanding the ideas behind an approximate method for dealing with load combinations. This method is very suitable for use in connection with the level 2 methods presented in chapters 5 and 6.

10.2 THE LOAD COMBINATION PROBLEM

One of the fundamental problems in dealing with time-varying loads modelled by stochastic processes is connected with estimation of the probability that the stochastic process defined as the sum of the individual processes crosses a given barrier (threshold) during the reference period T. More specifically, let two loads (or load effects) be modelled by stationary and independent stochastic processes $\{X_1(t), 0 \leq t \leq T\}$ and $\{X_2(t), 0 \leq t \leq T\}$. In the following all stochastic processes will have the same index set so that the shorter notations $\{X_1(t)\}$, $\{X_2(t)\}$, etc. and $\{X_1\}$, $\{X_2\}$, etc. can be used. The combination problem can then be formulated in the following way. What is the probability that the process

$$\{X(t)\} = \{X_1(t) + X_2(t)\} \tag{10.1}$$

has a value larger than $x(t) = \xi$ during the reference period $0 \leq t \leq T$? This probability is

$$P(\max X(t) > \xi,\ t \in [0; T]) =$$

$$= P(X(0) > \xi) + P(\text{one or more upcrossings of } \xi \mid X(0) < \xi) \tag{10.2}$$

where $P(X(0) > \xi)$ is the probability that the process $\{X(t)\}$ has a value greater than ξ at $t = 0$ and the last term in (10.2) is approximately equal to

$$P(\text{one or more upcrossings of } \xi) = \sum_{n=1}^{n=\infty} P(n \text{ upcrossings of } \xi) \tag{10.3}$$

The expected number of upcrossings (positive crossings) of a level ξ per unit time for a stationary process is denoted $E[N'_+(\xi)]$ in section 9.4. It is convenient to use a shorter notation $\nu_X(\xi)$ here. When $\nu_X(\xi) = E[N'_+(\xi)]$ is known, then the expected number of upcrossings in the time interval $[0; T]$ is equal to $\nu_X(\xi)\cdot T$, i.e.

$$E[\text{number of upcrossings}] = \nu_X(\xi)\cdot T = \sum_{n=1}^{n=\infty} n\cdot P(n \text{ upcrossings of } \xi) \tag{10.4}$$

It follows from (10.2), (10.3) and (10.4) that

$$P(\max X(t) > \xi, t \in [0; T]) \leqslant P(X(0) > \xi) + \nu_X(\xi)\cdot T \tag{10.5}$$

In general $P(X(0) > \xi) \ll \nu_X(\xi)\cdot T$ and for most practical reliability problems $\nu_X(\xi)\cdot T \ll 1$. In such cases $\nu_X(\xi)\cdot T$ is a good approximation of $P(\max X(t) > \xi, t \in [0; T])$, i.e.

$$P(\max (X(t) > \xi, t \in [0; T]) \approx \nu_X(\xi)\cdot T \tag{10.6}$$

The left hand side of (10.6) is equal to $1 - F_\Xi(\xi)$, where F_Ξ is the distribution function of the maximum value of the stochastic process $\{X(t)\}$ in the time interval $[0; T]$. Therefore,

$$F_\Xi(\xi) \approx 1 - \nu_X(\xi)\cdot T \tag{10.7}$$

where $\nu_X(\xi)\cdot T \ll 1$. By (10.7) the problem of calculating the distribution function F_Ξ for the maximum value of $\{X(t)\} = \{X_1(t) + X_2(t)\}$ is reduced to that of determining the rate of upcrossings (the expected number of positive crossings) $\nu_X(\xi) = E[N'_+(\xi)]$ for $\{X(t)\}$. Unfortunately, exact expressions for $\nu_X(\xi)$ are only known for some special kinds of processes. An obvious way of calculating $\nu_X(\xi)$ is to use Rice's formula (9.34)

$$\nu_X(\xi) = E[N'_+(\xi)] = \int_0^\infty \dot{x} f_{X\dot{X}}(\xi, \dot{x})d\dot{x} \tag{10.8}$$

where $f_{X\dot{X}}$ is the joint density function for the process $\{X(t)\}$ and its derivative process $\{\dot{X}(t)\}$. The joint density function $f_{X\dot{X}}$ can be derived by the so-called convolution integral

$$f_{X\dot{X}}(x, \dot{x}) = \int_{x_1=-\infty}^{\infty} \int_{\dot{x}_1=-\infty}^{\infty} f_{X_1\dot{X}_1}(x_1, \dot{x}_1) f_{X_2\dot{X}_2}(x - x_1, \dot{x} - \dot{x}_1) d\dot{x}_1 dx_1 \qquad (10.9)$$

where $f_{X_1\dot{X}_1}$ and $f_{X_2\dot{X}_2}$ are the joint density functions for X_1, $\dot{X}_1$ and X_2, $\dot{X}_2$, respectively. Note that equation (10.9) is a generalization of the well-known convolution integral in elementary probability theory. Also note that the first step in calculating $f_{X\dot{X}}$ is to calculate $f_{X_1X_1}$ and $f_{X_2\dot{X}_2}$. This is in general difficult, but it has been done for some special stochastic processes.

By inserting (10.9) in Rice's formula (10.8) one gets

$$\nu_X(\xi) = \int_{\dot{x}=0}^{\infty} \dot{x} \int_{x_1=-\infty}^{\infty} \int_{\dot{x}_1=-\infty}^{\infty} f_{X_1\dot{X}_1}(x_1, \dot{x}_1) f_{X_2\dot{X}_2}(\xi - x_1, \dot{x} - \dot{x}_1) d\dot{x}_1 dx_1 d\dot{x} \qquad (10.10)$$

(10.10) can be written in a more convenient form by the substitution $\dot{x} = \dot{x}_1 + \dot{x}_2$

$$\nu_X(\xi) = \int_{x=-\infty}^{\infty} \int_{\dot{x}_1=-\infty}^{\infty} \int_{\dot{x}_2=-\dot{x}_1}^{\infty} (\dot{x}_1 + \dot{x}_2) f_{X_1\dot{X}_1}(x, \dot{x}_1) f_{X_2\dot{X}_2}(\xi - x, \dot{x}_2) d\dot{x}_2 d\dot{x}_1 dx$$

$$= \int_{x=-\infty}^{\infty} \int_{\omega} \dot{x}_1 f_{X_1\dot{X}_1}(x, \dot{x}_1) f_{X_2\dot{X}_2}(\xi - x, \dot{x}_2) d\omega dx$$

$$+ \int_{x=-\infty}^{\infty} \int_{\omega} \dot{x}_2 f_{X_1\dot{X}_1}(x, \dot{x}_1) f_{X_2\dot{X}_2}(\xi - x, \dot{x}_2) d\omega dx \qquad (10.11)$$

where the domain ω in the $\dot{x}_1 \dot{x}_2$-plane is shown in figure 10.2.

In conclusion the procedure for evaluating the distribution function F_{Ξ} for the maximum value of the stochastic process $\{X(t)\} = \{X_1(t) + X_2(t)\}$ in the time interval $[0; T]$ is

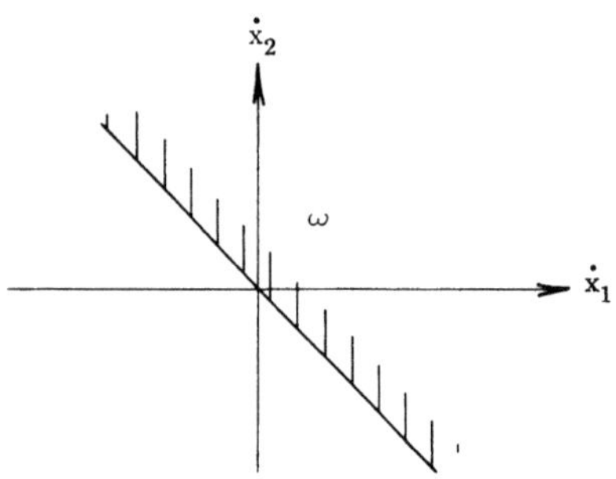

Figure 10.2

(1) Calculate $f_{X_1\dot{X}_1}$ and $f_{X_2\dot{X}_2}$ for the two processes $\{X_1\}$ and $\{X_2\}$

(2) Find $\nu_X(\xi) = E[N'_+(\xi)]$ by evaluating the integrals in (10.11)

(3) Find an approximate expression for F_{Ξ} from (10.7)

Step (2) above can only be performed exactly for special density functions. Usually numerical integration must be used. However, upper and lower bounds for $\nu_X(\xi)$ can be derived by changing the domain of integration (ω) in (10.11) in an appropriate way. The upper bound is especially useful so its derivation will be shown here.

The upper bound is obtained by changing the domain of integration in the first integral on the right hand side from ω to ω_1 and the domain of integration in the second integral on the right hand side from ω to ω_2, where ω_1 and ω_2 are shown in figure 10.3. Clearly, an upper bound of $\nu_X(\xi)$ is then obtained

$$\nu_X(\xi) \leqslant \int_{x=-\infty}^{\infty}\int_{\dot{x}_1=0}^{\infty}\int_{\dot{x}_2=-\infty}^{\infty} \dot{x}_1 f_{X_1\dot{X}_1}(x,\dot{x}_1) f_{X_2\dot{X}_2}(\xi-x,\dot{x}_2)\,d\dot{x}_2\,d\dot{x}_1\,dx$$

$$+\int_{x=-\infty}^{\infty}\int_{\dot{x}_1=-\infty}^{\infty}\int_{\dot{x}_2=0}^{\infty} \dot{x}_2 f_{X_1\dot{X}_1}(x,\dot{x}_1) f_{X_2\dot{X}_2}(\xi-x,\dot{x}_2)\,d\dot{x}_2\,d\dot{x}_1\,dx$$

$$=\int_{-\infty}^{\infty}\nu_{X_1}(x) f_{X_2}(\xi-x)\,dx + \int_{-\infty}^{\infty}\nu_{X_2}(\xi-x) f_{X_1}(x)\,dx \qquad (10.12)$$

where $\nu_{X_1}(x)$ and $\nu_{X_2}(\xi-x)$ are rates of upcrossings for the processes $\{X_1\}$ and $\{X_2\}$. The integrals in (10.12) are much more convenient than the integrals in (10.11) because they only involve rates of upcrossing of the processes $\{X_1\}$ and $\{X_2\}$ and the corresponding density functions. It has been shown in the literature (see the references at the end of chapter 10) that the upper bound (10.12) is very close to the exact result, so that it can be used as an approximation for $\nu_X(\xi)$.

$$\nu_X(\xi) \approx \int_{-\infty}^{\infty}\nu_{X_1}(x) f_{X_2}(\xi-x)\,dx + \int_{-\infty}^{\infty}\nu_{X_2}(\xi-x) f_{X_1}(x)\,dx \qquad (10.13)$$

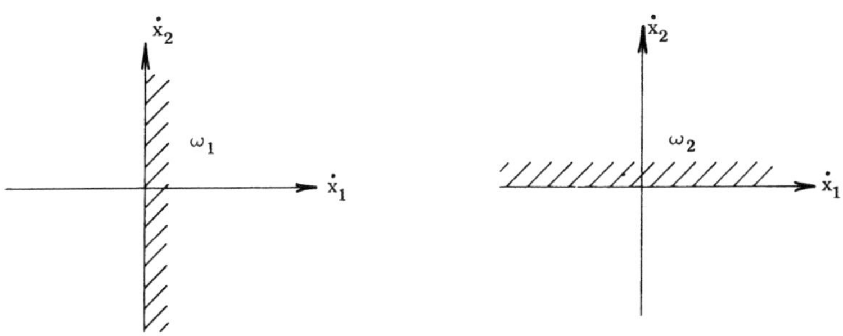

Figure 10.3

Only the sum of two independent processes has been treated above. However, the same procedure can be used for sums of three or more processes. For example, for the sum $\{X\}$ of three independent processes $\{X_1\}$, $\{X_2\}$ and $\{X_3\}$ one gets

$$\nu_X(\xi) \leq \int_{-\infty}^{\infty} [\nu_{X_1}(x) f_{X_2+X_3}(\xi - x) + \nu_{X_2}(x) f_{X_1+X_3}(\xi - x) + \nu_{X_3}(x) f_{X_1+X_2}(\xi - x)] dx \tag{10.14}$$

where the density functions $f_{X_i+X_j}$ are determined as usual by the convolution integral

$$f_{X_i+X_j}(x) = \int_{-\infty}^{\infty} f_{X_i}(t) f_{X_j}(x-t) dt \tag{10.15}$$

(10.14) can easily be generalized to sums of more than three independent processes.

10.3 THE FERRY BORGES - CASTANHETA LOAD MODEL

In this section a simple load model suggested by Ferry Borges and Castanheta will be presented. In this model real loading processes are greatly simplified in such a way that the mathematical problems connected with estimating the distribution function of the maximum value of a sum of loading processes are avoided. Further, the Ferry Borges - Castanheta load model is very suitable in connection with the level 2 methods presented in chapters 5 and 6.

For each load process $\{X_i\}$ it is assumed that the load changes after equal so-called *elementary intervals of time* τ_i. This is illustrated in figure 10.4, where the *reference period* T (e.g. 1 year) is divided into n_i intervals of equal length $\tau_i = T/n_i$. n_i is called the *repetition number*. Further it is assumed that the load is constant in each elementary interval. The loads in the elementary intervals are identically distributed and mutually independent random variables with a density function (point-in-time distribution) f_{X_i}. This density function is shown as a continuous density function in figure 10.4 but it can also be a density function of the mixed type (see page 22). This is convenient if, for example, it is desirable to have the load value 0 with a finite probability. Let the point-in-time distribution for load process $\{X_i\}$ be f_{X_i} and the corresponding distribution function F_{X_i} then the distribution of the maximum value in the reference period T is $(F_{X_i})^{n_i}$, i.e. (see (3.5))

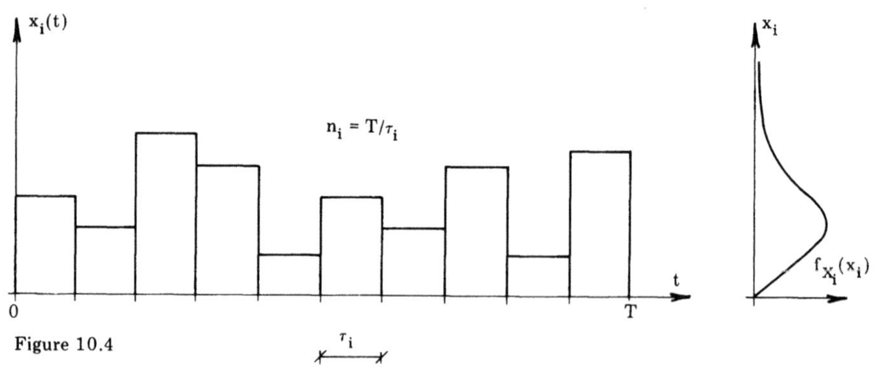

Figure 10.4

$$F_{\max_T X_i}(x_i) = (F_{X_i}(x_i))^{n_i} \tag{10.16}$$

Therefore, for this so-called *rectangular pulse process* it is a simple task to calculate the distribution of the maximum value in the reference period T.

When combinations of load processes $\{X_1\}, \{X_2\}, \ldots, \{X_r\}$ are considered it is assumed in the Ferry Borges - Castanheta load model that the loads are stochastically independent with integer repetition numbers n_i, where

$$n_1 \leqslant n_2 \leqslant \ldots \leqslant n_i \leqslant \ldots \leqslant n_r \tag{10.17}$$

and where (Z_+ is the set of positive natural numbers)

$$n_i/n_{i-1} \in Z_+ \quad \text{for } i \in \{2, 3, \ldots, r\} \tag{10.18}$$

The conditions (10.17) and (10.18) are illustrated in figure 10.5 where r = 3 and $n_1 = 2$, $n_2 = 6$ and $n_3 = 12$.

Although the Ferry Borges - Castanheta load model presented above is a gross simplification of the real loading situation, experiences seem to verify that the model is capable of reflecting the most important characteristics of load combinations.

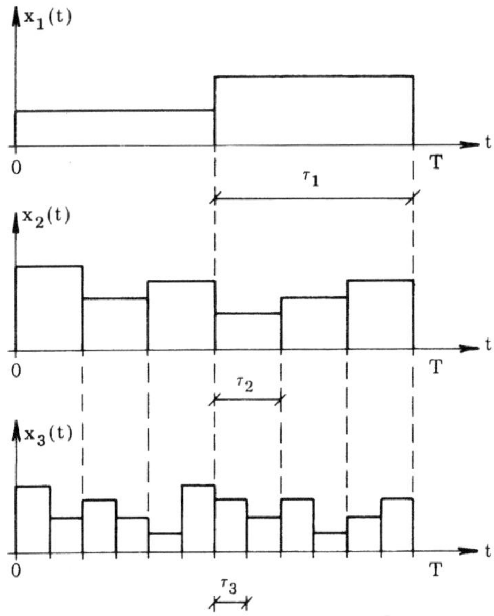

Figure 10.5

10.4 COMBINATION RULES

It has been emphasized earlier that two loading processes will usually not reach their maximum value in a given reference period T at the same instant of time. It is therefore too conservative to replace $\max_T \{X_1(t) + \ldots + X_r(t)\}$ by $\max_T \{X_1(t)\} + \ldots + \max_T \{X_2(t)\}$. On the other hand, $\max_T \{X_1(t) + \ldots + X_r(t)\}$ is a very complicated stochastic variable to use in practice, so some kind of approximation must be made.

Using *Turkstra's rule,* $\max_T \{X_1(t) + \ldots + X_r(t)\}$ is replaced by r stochastic variables, namely

$$\left.\begin{aligned} Z_1 &= \max_T \{X_1(t)\} + X_2(t^*) + \ldots + X_r(t^*) \\ Z_2 &= X_1(t^*) + \max_T \{X_2(t)\} + \ldots + X_r(t^*) \\ &\vdots \\ Z_r &= X_1(t^*) + X_2(t^*) + \ldots + \max_T \{X_r(t)\} \end{aligned}\right\} \qquad (10.19)$$

where t^* is an arbitrary point in time. By this rule the reliability of a structure is only checked at those points in time where the individual load processes reach their maximum value. Therefore, the reliability of a structure will be overestimated. However, it has been shown that this overestimation is usually very small.

A more refined rule has been formulated in connection with the Ferry Borges - Castanheta load model presented in section 10.3. In this model the loading processes $\{X_1\}, \{X_2\}, \ldots, \{X_r\}$ are rectangular load processes with $n_1, n_2, \ldots, n_r$ repetitions in the reference period T, where $n_1 \leq n_2 \leq \ldots \leq n_r$.

For $r = 2$, the rule gives the following 2 combinations for the loads:

Combination No.	No. of repetitions of load 1	2
1	n_1	n_2/n_1
2	1	n_2

Table 10.1

For $r = 3$, the rule gives the following 4 combinations:

Combination No.	No. of repetitions of load		
	1	2	3
1	n_1	n_2/n_1	n_3/n_2
2	1	n_2	n_3/n_2
3	n_1	1	n_3/n_1
4	1	1	n_3

Table 10.2

In general with r actions 2^{r-1} different combinations of load have to be considered.

Example 10.1. Let the number of rectangular pulse processes be r = 3 and let the number of repetitions be $n_1 = 3$, $n_2 = 6$ and $n_3 = 30$ in the reference period T. According to table 10.2 the following combinations have to be checked:

Combination 1: Max(3 rept. of X_1) + Max(2 rept. of X_2) + Max(5 rept. of X_3)

Combination 2: (1 rept. of X_1) + Max(6 rept. of X_2) + Max(5 rept. of X_3)

Combination 3: Max(3 rept. of X_1) + (1 rept. of X_2) + Max(10 rept. of X_3)

Combination 4: (1 rept. of X_1) + (1 rept. of X_2) + Max(30 rept. of X_3)

Examples 10.2 and 10.3 show how the reliability index β can be calculated for a structure loaded by r = 2 time-varying loads modelled by Gaussian rectangular pulse processes.

Example 10.2. Consider the indeterminate beam shown in figure 10.6 with two time-dependent loads $p_1(t)$ and $p_2(t)$. Let $p(t) = p_1(t) + p_2(t)$. Let $p_1(t)$ be a realisation of a stochastic process $\{P_1(t)\}$ and $p_2(t)$ of a stochastic process $\{P_2(t)\}$. Further, let $\{P_1(t)\}$ be a Gaussian pulse process with $\mu_{P_1(t)} = 3$ kN and $\sigma_{P_1(t)} = 0.3$ kN and with $n_1 = 1$ repetitions in the design life (reference period) T = 1 year. Likewise, let $\{P_2(t)\}$ be a Gaussian pulse process with $\mu_{P_2(t)} = 2$ kN and $\sigma_{P_2(t)} = 0.2$ kN but with $n_2 = 12$ repetitions in the design life 1 year. Realisations of the pulse processes are shown in figure 10.7.

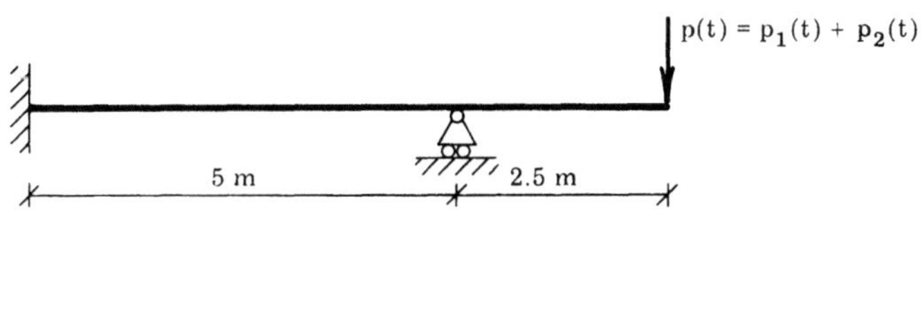

Figure 10.6

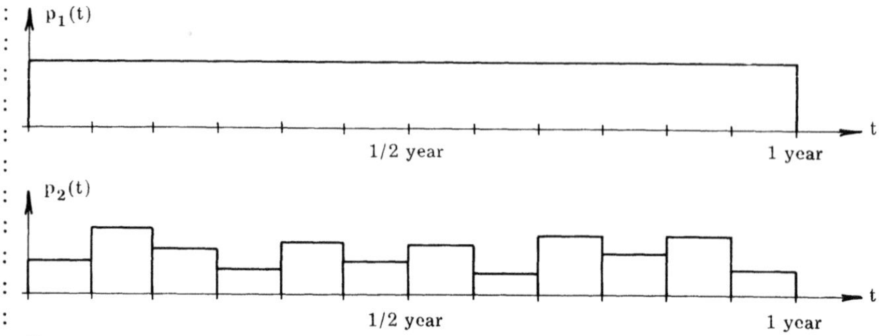

Figure 10.7

The maximum value $P_{2,max}$ of the load process $\{P_2(t)\}$ is then

$$P_{2,max} = \max_{i=1,\dots,12} [P_{2i}] \tag{10.20}$$

where P_{2i} is the load level in pulse i. Due to the independence of the pulses and their identical Gaussian distributions the distribution function F_{X_2} for $X_2 = P_{2,max}$ is given by

$$F_{X_2}(x_2) = \Phi^{12}\left(\frac{x_2 - \mu_{P_2}}{\sigma_{P_2}}\right) \tag{10.21}$$

Note that X_2 is not Gaussian distributed. Therefore, in connection with level 2 reliability analysis or design a transformation must be performed, for example as shown in section 6.4. By this transformation the distribution of $X_2 = P_{2,max}$ is replaced by a normal distribution with mean μ'_{X_2} and standard deviation σ'_{X_2}, where (see (6.37) and (6.38))

$$\sigma'_{X_2} = \frac{\varphi\left(\Phi^{-1}\left(\Phi^{12}\left(\frac{x_2^\star - 2}{0.2}\right)\right)\right)}{12\cdot\Phi^{11}\left(\frac{x_2^\star - 2}{0.2}\right)\varphi\left(\frac{x_2^\star - 2}{0.2}\right)}\cdot 0.20 \tag{10.22}$$

$$\mu'_{X_2} = x_2^\star - \Phi^{-1}\left(\Phi^{12}\left(\frac{x_2^\star - 2}{0.2}\right)\right)\sigma'_{X_2} \tag{10.23}$$

$x_2^\star$ is the x_2-coordinate for the design point.

Example 10.3. Consider the same beam as in example 10.2 and with the same loads. Further, let the safety margin M be given by

$$M = M_F - \frac{5}{2}\left(P_1 + \max_{i=1,\dots,12}[P_{2i}]\right) \tag{10.24}$$

where the critical limit moment M_F is a normally distributed random variable with μ_{M_F} = 20 kNm and σ_{M_F} = 2 kNm. Introduce the random variable $X_1 = M_F - \frac{5}{2}P_1$. X_1 is normally distributed with

$$\mu_{X_1} = 20 - 7.5 = 12.5 \text{ kNm} \tag{10.25}$$

$$\sigma_{X_1} = \sqrt{2^2 + \left(\frac{5}{2}\cdot 0.3\right)^2} = 2.14 \text{ kNm} \tag{10.26}$$

The safety margin can then be reformulated

$$M = X_1 - \frac{5}{2} X_2 \tag{10.27}$$

In the normalised coordinate system the failure surface is then given by

$$(12.5 + 2.14\, x_1) - 2.5(\mu'_{X_2} + \sigma'_{X_2} x_2) = 0 \tag{10.28}$$

The reliability index β can now be calculated by the same iterative technique as used in example 6.8. With the usual notation

$$\beta = \frac{2.5\,\mu'_{X_2} - 12.5}{2.14\,\alpha_1 - 2.5\,\sigma'_{X_2}\alpha_2} \tag{10.29}$$

$$\alpha_1 = -\frac{1}{k} \cdot 2.14 \tag{10.30}$$

$$\alpha_2 = +\frac{1}{k} \cdot 2.5\,\sigma'_{X_2} \tag{10.31}$$

where σ'_{X_2} and μ'_{X_2} are given by (10.22) and (10.23) with

$$\frac{x_2^\star - 2}{0.2} = \beta\alpha_2 \tag{10.32}$$

	Start	Iteration No.		
		1	2	3
β	3.00	3.87	3.11	3.11
α_1	−0.717	−0.989	−0.990	−0.991
α_2	0.717	0.148	0.141	0.138
$\frac{(x_2^\star - 2)}{0.2}$	2.151	0.573	0.439	0.429
σ'_{X_2}	0.128	0.121	0.119	0.119
μ'_{X_2}	2.31	2.31	2.32	2.32

Table 10.3. The reliability index is $\beta = 3.11$.

Exercise 10.2. Show that the reliability index for the structure in example 10.3 is $\beta = 3.19$ if the number of repetitions n_2 is equal to 6 (and not 12) but with all other data unchanged.

Example 10.4. Consider the same structure as in examples 10.2 and 10.3. The variation of the reliability index β with the number of repetitions n_2 for the load process $\{P_2(t)\}$ is shown in figure 10.8.

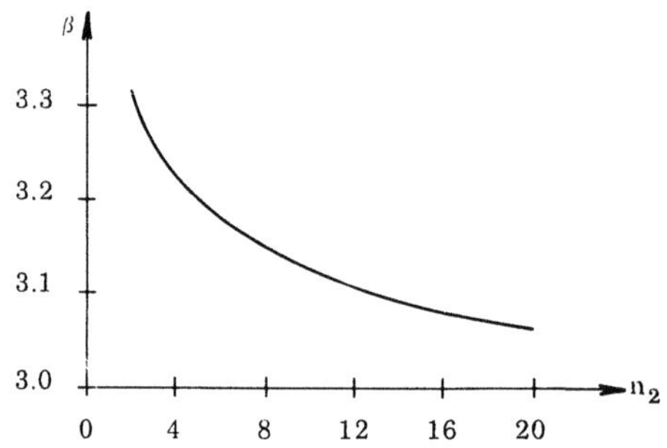

Figure 10.8

Example 10.5. Consider again the structure analysed in examples 10.2 and 10.3, but now the structure has to be designed so that it has a reliability index $\beta = 4.00$. Let the critical limit moment M_F be normally distributed with unknown mean value μ_{M_F} and standard deviation $\sigma_{M_F} = 0.1 \cdot \mu_{M_F}$. All other data are unchanged.

The safety margin with $X_1 = M_F$, $X_2 = P_1$ and $X_3 = P_{2,max}$ is

$$M = X_1 - 2.5\,X_2 - 2.5\,X_3 \tag{10.33}$$

where X_1 is $N(\mu_{M_F}, 0.1\,\mu_{M_F})$, X_2 is $N(3, 0.3)$ and X_3 is $\approx N(\mu'_{X_3}, \sigma'_{X_3})$. The formulae for the iterative process are

$$\mu_{M_F} = \frac{7.5 + 3\,\alpha_2 + 2.5\mu'_{X_3} + 10\sigma'_{X_3}\alpha_3}{1 + 0.4\,\alpha_1} \tag{10.34}$$

$$\alpha_1 = -\frac{1}{k}\mu_{M_F}\,0.1 \tag{10.35}$$

$$\alpha_2 = \frac{1}{k}\,0.75 \tag{10.36}$$

$$\alpha_3 = \frac{1}{k}\,2.5\,\sigma'_{X_3} \tag{10.37}$$

and the iteration scheme is:

	Start	Iteration No.		
		1	2	3
μ_{M_F}		20.5	23.0	23.1
α_1	−0.577	−0.928	−0.944	−0.944
α_2	0.577	0.340	0.307	0.307
α_3	0.577	0.152	0.124	0.123
σ'_{X_3}	0.134	0.122	0.120	0.120
μ'_{X_3}	2.30	2.31	2.32	2.32

Table 10.4. With $\mu_{M_F} = 23.1$ kNm the reliability index $\beta = 4.00$.

In the final example 10.6 it is shown how the reliability index β can be calculated for a structure with 3 time-varying loads modelled as Gaussian pulse processes.

Example 10.6. Consider the simply supported beam shown in figure 10.9. The beam is loaded by 3 uniformly distributed time-dependent loads $p_1(t)$, $p_2(t)$ and $p_3(t)$. The Ferry Borges - Castanheta load modelling is used for the corresponding load processes $\{P_1(t)\}$, $\{P_2(t)\}$ and $\{P_3(t)\}$. The modelling data are shown in table 10.3. The safety margin is

$$M = M_F - \frac{1}{2} \cdot 25(P_1 + \max_{i=1,\ldots,6} [P_{2i}] + \max_{i=1,\ldots,180} [P_{3i}]) \tag{10.38}$$

where the critical limit moment M_F is assumed to be N(12.50 kNm, 1.25 kNm). Note that in the last term in (10.38) the number of elementary intervals is only 180 due to the fact that the load process $\{P_2(t)\}$ is only assumed to be active for 1/2 year with $n_2 = 6$ elementary intervals.

Equation (10.38) can be rewritten

$$M = X_1 - \frac{25}{8} \max_{i=1,\ldots,6} [X_{2i} + \max_{i=1,\ldots,30} [X_{3i}]] \tag{10.39}$$

where

$X_1 = M_F - \frac{25}{8} P_1$ is N(10.9375, 1.3975)

X_{2i} is N(− 0.20, 0.40)

X_{3i} is N(− 2.00, 1.00)

Load process	No. of repetitions	$\mu_{P_i(t)}$, kN	$\sigma_{P_i(t)}$, kN	Distribution
$\{P_1(t)\}$	n_1 = 1/year	0.50	0.20	Gaussian
$\{P_2(t)\}$	$n_2 = 6/\frac{1}{2}$ year	−0.20	0.40	Gaussian
$\{P_3(t)\}$	n_3 = 360/year	−2.00	1.00	Gaussian

Table 10.5

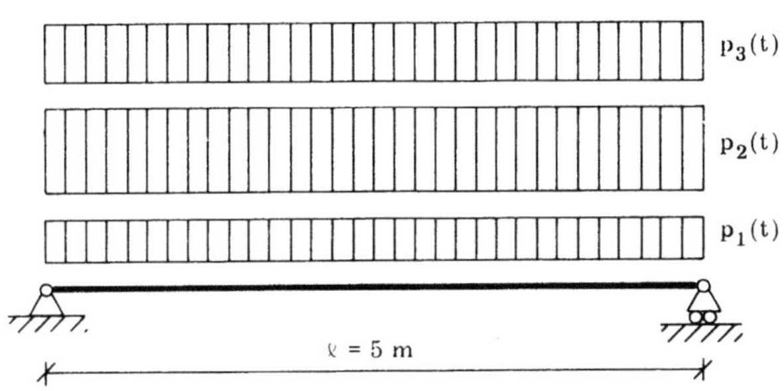

Figure 10.9

Let $X_4 = \max_{i=1,\dots,30} [X_{3i}]$ be approximated by a normal distribution $N(\mu'_{X_3}, \sigma'_{X_3})$ and let $X_5 = X_2 + X_4$. X_5 will then be normally distributed $N(-0.20 + \mu'_{X_3}, \sqrt{0.40^2 + (\sigma'_{X_3})^2})$.

Finally, let $X_6 = \max_{i=1,\dots,6} [X_{5i}]$ be approximated by a normal distribution $N(\mu'_{X_5}, \sigma'_{X_5})$.

All approximations are made at the design point $\mu'_{X_i} + x_i\,\sigma'_{X_i}$. In the normalised coordinate system the failure surface is then given by

$$(10.935 + 1.3975\,x_1) - \frac{25}{8}(\mu'_{X_5} + \sigma'_{X_5} x_6) = 0 \tag{10.40}$$

and the iteration formulas are

$$\beta = \frac{\frac{25}{8}\mu'_{X_5} - 10.935}{1.3975\,\alpha_1 - \frac{25}{8}\sigma'_{X_5}\alpha_6} \tag{10.41}$$

$$\left.\begin{aligned} \alpha_1 &= -\frac{1}{k_1}1.3975 , & x_1 &= \alpha_1\beta \\ \alpha_6 &= \frac{1}{k_1}\frac{25}{8}\sigma'_{X_5} , & x_6 &= x_5 = \alpha_6\beta \end{aligned}\right\} \tag{10.42}$$

$$\left.\begin{aligned} \alpha_2 &= \frac{1}{k_2}0.40 , & x_2 &= \alpha_2\beta^\star \\ \alpha_4 &= \frac{1}{k_2}\sigma'_{X_3} , & x_3 &= x_4 = \alpha_4\beta^\star \end{aligned}\right\} \tag{10.43}$$

$$\beta^\star = \frac{\mu'_{X_5} + 0.20 - \mu'_{X_3} + \beta\alpha_6\sigma'_{X_5}}{\sqrt{0.40^2 + (\sigma'_{X_3})^2}} \tag{10.44}$$

The iteration scheme is as follows:

	Start	Iteration No.				
		1	2	3	4	5
σ'_{X_3}	1.000	0.503	0.513	0.701	0.783	0.811
μ'_{X_3}	−2.000	−0.005	−0.008	−0.233	−0.435	−0.521
σ'_{X_5}	1.077	0.358	0.541	0.698	0.777	0.802
μ'_{X_5}	−2.200	0.557	0.442	0.302	0.121	0.045
$\beta^\star$	3.000	0.450	2.951	3.386	3.662	3.738
α_1	−0.707	−0.781	−0.637	−0.539	−0.499	−0.487
α_6	0.707	0.625	0.771	0.842	0.867	0.874
α_2	0.707	0.622	0.615	0.496	0.455	0.442
α_4	0.707	0.783	0.789	0.868	0.891	0.897
β	3.00	5.17	4.45	3.88	3.77	3.76

Table 10.6

: It is important to note that the values for $\mu_{P_i(t)}$ and $\sigma_{P_i(t)}$ in table 10.5 are values adjusted
: in such a way that the approximated normal distributions for the maximum distributions
: are acceptable.

BIBLIOGRAPHY

[10.1] Ferry Borges, J. & Castanheta, M.: *Structural Safety*. 2nd edition. Laboratorio Nacional de Engenharia Civil, Lisbon, 1972.

[10.2] Madsen, H. O.: *Load Models and Load Combinations.* Thesis. Technical University of Denmark, Lyngby, February, 1979.

[10.3] Madsen, H. O.: *Some Experience with the Rackwitz-Fiessler Algorithm for the Calculation of Structural Reliability under Combined Loading.* DIALOG 77. Danish Engineering Academy, Lyngby, pp. 73-98.

[10.4] Madsen, H., Kilcup, R. & Cornell, C. A.: *Mean Upcrossing Rates for Sums of Pulse-Type Stochastic Load Processes.* Proc. Specialty Conf. on Probabilistic Mech. and Struct. Reliability, ASCE, Tucson, Arizona, January 1979.

[10.5] Nordic Committee on Building Regulations: *Recommendation for Loading and Safety Regulations for Structural Design.* NKB-Report No. 36, Nov. 1978.

[10.6] Rackwitz, R. & Fiessler, B.: *Two Applications of First Order Reliability Theory for Time-variant Loadings.* LKI, Heft 17, Technische Universität München, 1977.

[10.7] Turkstra, C. J. & Madsen, H. O.: *Load Combinations in Codified Structural Design.* J. Struct. Div., ASCE, Vol. 106, No. St. 12, December 1980.

[10.8] Turkstra, C. J.: *Application of Bayesian Decision Theory. Study No. 3: Structural Reliability and Codified Design.* Solid Mechanics Division, University of Waterloo, Waterloo, Canada, 1970.

[10.9] Wen, Yi-Kwei: *Statistical Combination of Extreme Loads.* J. Struct. Div., ASCE, Vol. 103, No. ST5, May 1977, pp. 1079 - 1093.

Chapter 11

APPLICATIONS TO STRUCTURAL CODES

11.1 INTRODUCTION

Structural codes are documents which, by their very nature, are subject to periodic revision and amendment, but the decade 1970 - 80 was a time of marked activity in code development. This is still continuing. The main features have been

- the replacement of many simple design rules by more scientifically-based calculations derived from experimental and theoretical research,
- the move towards limit state design - whereby the designer and/or code writer specifies the relevant performance requirements (limit states) for each structure explicitly; and where separate sets of calculations are required to check that the structure will not attain each limit state (at a given level of probability),
- the replacement of single safety factors or load factors by sets of partial coefficients,
- the improvement of rules for the treatment of combinations of loads and other actions,
- the use of structural reliability theory in determining rational sets of partial coefficients, and
- the preparation of model codes [11.7] for different types of structural materials and forms of construction; and steps towards international code harmonisation, particularly within the European Economic Community (EEC).

It should not be thought that all these developments have been fully co-ordinated, or that all the changes to practical codes that have taken place are necessarily of great benefit. Indeed many recent changes in structural codes have not been met with enthusiasm by practising engineers, often for good reason. Nevertheless, each of the features mentioned above is of relevance to future code development.

In comparison with the idealised models used for calculation purposes, the *actual* behaviour of most structures is extremely complex and there is a tendency, as more research is undertaken and more becomes known, for the design procedures set out in structural codes to become increasingly lengthy and involved. Such changes generally increase design costs and increase the risk of major errors being made. They cannot be classėd as improvements unless the new procedures result in improved standards of safety and/or reduced costs of construction and maintenance.

It is therefore clear that the »best» codes are not necessarily those with the most scientifically advanced design clauses. As will be discussed later, there may often be advantages in using simplified design rules. The effect of this will be to make the overall construction slightly less economic and the reliability of those structures designed to the code marginally more variable, for any specified standard of reliability.

In previous chapters, various aspects of structural reliability theory have been discussed, together with the problems of modelling load and resistance variables. In this chapter we consider how these techniques can be used in the development of conventional structural codes.

11.2 STRUCTURAL SAFETY AND LEVEL 1 CODES

As mentioned in chapter 1, *level 1* design methods were described as »design methods in which appropriate degrees of structural reliability are provided on a structural element basis (occasionally on a structural basis) by the use of a number of partial safety factors (partial coefficients) related to pre-defined characteristic or nominal values of the major structural and loading variables». A level 1 code is therefore a conventional deterministic code in which the nominal strengths of the structural members designed to that code are governed by a number of *partial coefficients* or by equivalent means.

The safety and serviceability of practical structures are achieved by the use of suitable partial coefficients in design, together with appropriate *control measures*. Both are essential and it is helpful to distinguish their individual roles.

Let us first examine the role of partial coefficients. Consider a structure subjected to a random time-varying load Q having a specified nominal magnitude q_{sp}. The structure is proportioned to carry a design load $Q_d = \gamma_Q q_{sp}$, where γ_Q is a partial coefficient on live load. The effects of *increasing* γ_Q by, say, 20% will in general be

- an increase in the nominal capacity of the structure to support the load Q,
- an increase in the actual capacity of the structure to support the load Q,
- an increase in the sizes of the structural members and the self-weight of the structure,
- an increase in the cost of the structural system,
- some increase in the actual capacity of the structure to resist any other load Q', and
- an increase in the safety of the structure as characterised by a reduction in the probability that it will fail in any given reference period T.

If the design strength of a material is given by $e_d = e_{sp}/\gamma_m$, where e_{sp} is the specified material strength and γ_m is a partial coefficient, an increase in γ_m will in general have the same effects as an increase in γ_Q.

There are some circumstances, however, when increases in γ_Q or in γ_m may not give rise to these effects. For example, the actual load-carrying capacity of a structural member, as opposed to its nominal capacity, may decrease or may not significantly increase if, for example, any change in γ_Q or γ_m results in the designer using larger diameter reinforcing bars which, in spite of having the same specified yield stress as the bars they are replacing, may have a lower mean yield stress (see figure 3.9). Similarly, small changes in γ_Q or γ_m may sometimes have no effect on either the dimensions or the safety of some structural members. This is because of the discrete nature of many structural components (e.g. rolled steel beams) and the need to round up to the next section size above when designing. In such cases the actual strength, and hence the reliability, is not a continuous function of the partial coefficients.

We now consider the reasons for using partial coefficients as opposed to single safety factors or load factors. The main reason is that only by using partial coefficients can reasonably consistent standards of reliability be achieved over a range of different designs within any one code. As will be discussed in section 11.4, the most consistent standards can be achieved by associating a partial coefficient or some other safety element with each major source of uncertainty (i.e. with each basic variable). Partial coefficients are also essential for the rational treatment of load combinations, and in particular for situations in which the total load effect in part of a structure is the *difference* of two load effects of approximately similar magnitude but originating from different load sources - e.g. the effects of gravity loads and wind loads in the up-wind columns of a tall building.

We now return to the question of *control measures.* The safety and serviceability of a structure are influenced as much,if not more, by the nature of the control measures that are in operation as by the magnitude of the partial coefficients that are used in design. *Control* takes two main forms

- *quality control* of materials and fabrication, and
- controls to avoid the occurrence of major or *gross errors* in the design and construction processes.

Control of the first type is aimed at reducing variability in the mechanical properties of structural materials and maintaining appropriate mean properties. For example, the variability in the yield stress of steel can be reduced by improved control on chemical composition and rolling conditions. Such control will, in general, reduce the probability of structural failure and thus increase safety. Both the form and the parameters of the probabilistic models for resistance variables discussed in chapter 3 are dependent on the standards of quality control and inspection that are in operation.

Control of the second type is clearly more difficult to achieve since the sources of possible errors are almost unlimited. This is the subject of chapter 13.

We continue here with the problem of devising a suitable procedure for evaluating partial coefficients or other *safety elements* for a level 1 code. The term safety element is used as a generic term for *partial coefficients* and *additive safety elements* (see section 11.3.3). A logical sequence of steps is as follows

- set limits on the range of structures and materials for which the code will be applicable,
- specify the deterministic functional relationships to be used as the basis for each design clause,
- select the general form of the probabilistic models for the various load and resistance variables and model uncertainties,

- specify appropriate quality control measures and acceptance criteria for the manufacture and fabrication of basic materials and components,
- determine the parameters of the relevant models from loading data and from materials data obtained under the specified standards of quality control and inspection,
- select a suitable safety format - the number of partial coefficients and their position in the design equations (i.e. the variables associated with partial coefficients), etc.,
- select appropriate representative values of all basic random variables (e.g. nominal, characteristic or mean values) to be used as fixed deterministic quantities in the code,
- determine the magnitude of the partial coefficients to be used in conjunction with the above representative values to achieve the required standards of reliability.

Procedures such as this have already been used in the application of structural reliability theory to practical level 1 codes, e.g. [11.6], [11.10], [11.12]. Some of these steps have already been considered in some detail, e.g. the modelling of load and resistance variables, and others, e.g. quality control procedures, are beyond the scope of this book. In the remainder of this chapter we shall concentrate on the question of choosing suitable safety formats for structural codes and on the calculation of partial coefficients.

11.3 RECOMMENDED SAFETY FORMATS FOR LEVEL 1 CODES

The *safety format* of a code is defined as the way in which the various clauses of the code regulate the degree of safety, or more generally the reliability, of structures designed to the code. In particular, it concerns: the number of partial coefficients or other safety elements to be used, their positions in the design equations, and rules for load combinations.

The following recommendations for level 1 codes are based on the work of the International Joint Committee on Structural Safety [11.7], [11.8], and are likely to form the basis of a new international standard to replace ISO 2394: General principles for the verification of the safety of structures.

11.3.1 Limit state functions and checking equations

As discussed in chapters 4 and 5, the general conditions for a limit state not to be exceeded may be expressed as

$$f(X_1, X_2, \ldots, X_n) = f(\overline{X}) > 0 \tag{11.1}$$

where

$\overline{X}$ are the n basic random variables which influence the limit state, and

f is the limit state function (failure function).

The variables $\overline{X}$ may be sub-divided into variable loads and actions $\overline{Q}$, permanent loads $\overline{G}$, material properties $\overline{E}$, geometrical parameters $\overline{D}$, and model uncertainties $\overline{X}_m$ (see equation (1.1)). In addition, each limit state function is likely to involve one or more constants $\overline{c}$. Equation (11.1) may therefore be re-written as

$$f(\overline{Q}, \overline{G}, \overline{E}, \overline{D}, \overline{X}_m, \overline{c}) > 0 \tag{11.2}$$

For the purposes of a level 1 code, the equivalent deterministic criterion for *safety checking* (i.e. checking the sufficiency of a structure or structural member whose design properties are given) is

$$f(\overline{q}_d, \overline{g}_d, \overline{e}_d, \overline{d}_d, \overline{x}_{md}, \overline{c}) > 0 \tag{11.3}$$

where

f is the same limit state function as above, involving n quantities $\overline{x}_d$ and m constants $\overline{c}$, and

q_d is the deterministic *design value* of the random variable Q, etc.

If the aim is to *design*, as opposed to check, a particular structural member, it may often be possible to invert equation (11.3) to give the minimum design value of some convenient resistance variable - for example, a dimension D' or a section modulus, e.g.

$$d'_d = f'(\overline{q}_d, \overline{g}_d, \overline{e}_d, \overline{d}_d, \overline{x}_{md}, \overline{c}) \tag{11.4}$$

where

f' is a function related to f, involving (n — 1) quantities $\overline{x}_d$ and m constants $\overline{c}$.

Hence, the process of designing a structural member involves

- determination of the design loads $\overline{q}_d$,
- selection of materials and determination of the design values of their relevant mechanical properties $\overline{e}_d$,
- selection of primary dimensions $\overline{d}_d$ to satisfy the particular engineering and architectural requirements, and
- determination of the remaining unknown d'_d to satisfy equation (11.4).

In many cases it may not be possible or convenient to express equation (11.4) in explicit form in which case the design process will involve a number of trial-and-error calculations to find the minimum value of d' that satisfies the inequality (11.3). This will be recognised as the normal approach to design.

Let us now re-examine equation (11.2). For many structures it is possible to re-write this as

$$X_R \, r(\overline{E}, \overline{D}_R, \overline{c}) - X_S \, s(\overline{Q}, \overline{G}, \overline{D}_S, \overline{c}) > 0 \tag{11.5}$$

where

r represents a resistance function and $R = r(\cdot)$,

s represents a load effect or action effect function and $S = s(\cdot)$,

X_R is a model uncertainty associated with the particular form of the resistance function,

X_S is a model uncertainty associated with the particular form of the load effect or action effect function,

and where $\overline{D}_R$ and $\overline{D}_S$ are sets of different dimensions.

In equation (11.5) the resistance function r and the load effect function s are shown as uncoupled; and because they share no common variables the two terms are also statistically independent. If such uncoupling is possible, then the deterministic *checking equation* corresponding to equation (11.3) may be expressed as

$$\frac{1}{\gamma_R} x_{Rd}\, r(\overline{e}_d, \overline{d}_{Rd}, \overline{c}) - \gamma_S x_{Sd}\, s(\overline{q}_d, \overline{g}_d, \overline{d}_{Sd}, \overline{c}) > 0 \tag{11.6}$$

where

γ_R is a partial coefficient on the computed resistance

γ_S is a partial coefficient on the computed load effect

and where the subscript d denotes the *design value* of the variable.

The design process generally involves iterative or trial-and-error calculations to find a set of dimensions $\overline{d}_{Rd}$ which in conjunction with the design values of the load and strength variables satisfies the checking equation.

Equation (11.6) is the most general form of the checking equation for a structure in which R and S can be uncoupled. In this case, the safety or serviceability of a structure (the probability that the limit state defined by the particular form of the functions r and s will not be reached) can clearly be increased or decreased by adjusting *any or all* of the $(n-1)$ independent design values $\overline{x}_d$ (e.g. $\overline{q}_d$ or $\overline{e}_d$) and the two partial coefficients γ_R and γ_S. Substituting these values into equation (11.6) gives the required value of the remaining quantity - generally a dimension. Because there is an infinite number of sets of $(n-1)$ values $\overline{x}_d$ which will give the same design, the problem facing the code writer is to select the »best» set of values $\overline{x}_d$. This is discussed in section 11.4.

It should be noted that in practice the quantities R and S may often be correlated beacuse of common parameters. For example, the self-weight of a reinforced concrete beam and hence the mid-span bending moment S will be weakly correlated with the beam's moment-carrying capacity R, as both are functions of beam depth.

11.3.2 Characteristic values of basic variables

The term *characteristic value* was introduced in the late 1950's at the time when probabilistic concepts were first being introduced into structural codes; and when it was recognised that few basic variables have clearly defined upper or lower limits that can sensibly be used in design. Characteristic values of actions and material properties based on a prescribed probability p of not being exceeded were considered to be more rational than arbitrary selected values.

The characteristic value x_k of a basic random variable X is defined as the pth fractile of X given by

$$x_k = F_X^{-1}(p) \tag{11.7}$$

where

F_X^{-1} is the inverse distribution function of X, and

p is a probability which depends on the type of variable being considered (i.e. a load or a strength).

The selection of the probability p is to a large extent arbitrary but is influenced by the following considerations

- characteristic values of loads and other actions are values which should rarely be exceeded,
- characteristic values of material strength properties should normally be exceeded by actual properties,
- the values of p should neither be so large nor so small that the values x_k are not occasionally encountered,
- it is often sensible to use previously adopted nominal values as *specified characteristic values*, x_{sp}.

The distinction between *characteristic value* and *specified characteristic value* (specified value) should be made clear. The former is a fractile of a random variable, whereas the latter is some specified single value of the same quantity - a constant. For practical reasons it is generally necessary for the *user* of a level 1 code to work with specified values of all the design variables rather than with actual characteristic values, some of which will not be known at the design stage. For example, the actual characteristic value of the 28-day cube or cylinder strength of concrete is likely to depend on the particular supplier or contractor and is not known in advance. In this case it is necessary for the quality control procedures specified by the *code writers* to be such that the actual characteristic strength of the material exceeds the specified strength by an appropriate margin or with a stated probability. Similarly, the user of a code should normally work with specified deterministic values of loads and other actions; it is the responsibility of the code writers to relate these values to the distributions of the actual loads and actions, and to recommend associated partial coefficients or other safety elements.

11.3.3 Treatment of geometrical variables

Geometrical variables are of two main types - structural dimensions (e.g. the depth of a beam) and geometrical imperfections (e.g. the out-of-straightness of a column).

Structural dimensions: The uncertainties in most structural dimensions D are generally small and for this reason the mean value μ_D may be taken as the characteristic value (i.e. $d_k = \mu_D$). Tolerance limits are specified in codes for most structural dimensions, and if these are of the form

$$d_{sp} - \epsilon \leqslant D \leqslant d_{sp} + \epsilon \tag{11.8}$$

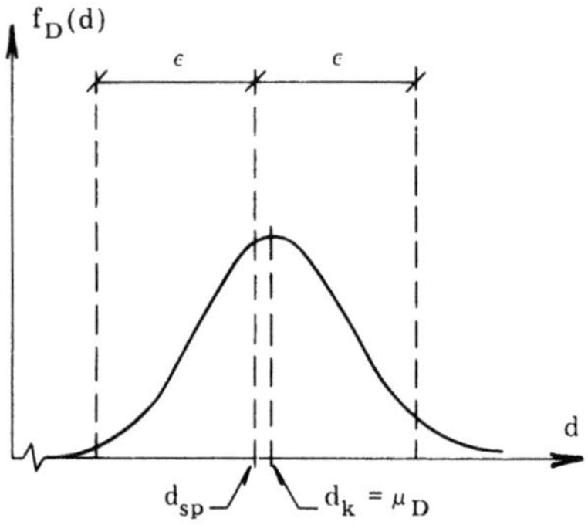

Figure 11.1 (a)

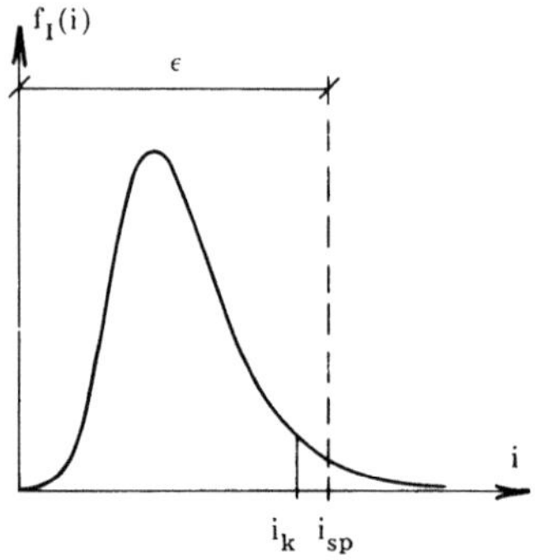

Figure 11.1 (b)

then the actual characteristic value d_k and the specified nominal value d_{sp} will generally be very close - see figure 11.1 (a). It should be noted, however, that unless the standard of inspection is high the probability that the dimension D will exceed the specified tolerance may not be negligible. See, for example, figure 3.6.

Geometrical imperfections: The strength of many structural members, for example most plates, columns and shell structures, depends not only on cross-sectional and overall dimensions but also on the magnitude of relevant geometrical imperfections I.

For such structures it is normal to specify an upper limit ϵ on the imperfection magnitude, i.e.

$$0 \leqslant I \leqslant \epsilon \tag{11.9}$$

In this case, ϵ can be taken as the specified characteristic value of I, i_{sp}. The probability that i_{sp} will be exceeded will generally be small and will depend on the standard of inspection. The actual characteristic value of the imperfection i_k can conveniently be chosen as the 95% fractile of I and the acceptance criteria designed so that i_{sp} exceeds i_k by an appropriate margin (or with a stated probability) - see figure 11.1 **(b).**

A histogram of some typical plate-panel imperfections (plate-panel out-of-flatness) obtained from measurements on the steel box-girder bridge at Aust in the U.K. is shown in figure 11.2. The quantity $\emptyset$ is the ratio of measured imperfection to the specified maximum imperfection ϵ.

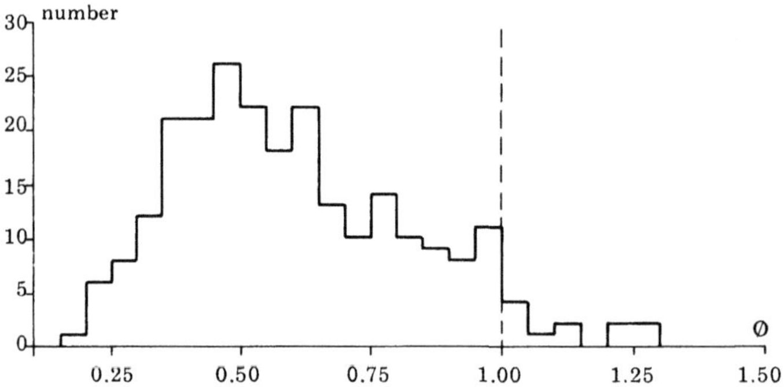

Figure 11.2 Plate panel imperfections - Aust Bridge.

Design values of dimensions and imperfections: Typically, the standard deviations of geometrical variables are independent of nominal dimensions (e.g. for given site conditions, the standard deviation in the thickness of a 100 mm slab is likely to be about the same as that of a 200 mm slab; giving a reduction in the coefficient of variation for increasing nominal thickness). For this reason the most uniform standards of reliability can be obtained over a range of different structures by using design values d_d and i_d of the geometrical variables related to the specified values, as follows

$$d_d = d_{sp} \pm \Delta_d \tag{11.10}$$

$$i_d = i_{sp} + \Delta_i \tag{11.11}$$

where Δ_d and Δ_i are *additive safety elements.*

For many structures, however, the probability of failure is insensitive to small variations in structural dimensions. For these cases, Δ_d and Δ_i should be set to zero and the uncertainties in D and I should be allowed for by modifications to the partial coefficients on the other design variables. A formal method for doing this is discussed in section 11.4.

11.3.4 Treatment of material properties

We shall restrict our attention to the strength properties of structural materials, denoted E. For each variable, the characteristic value e_k should be such that it has a reasonably high probability q (= 1 − p) of being exceeded in any single trial or test. Typically, q is taken to be between 0.95 and 0.99, corresponding to the 5% and 1% fractiles of the variable E. However, as mentioned in section 11.3.2, the user of a level 1 code may often not know the actual characteristic values for his material properties in advance, and it is generally necessary to design using specified characteristic values, e_{sp}. The acceptance criteria for a material should be devised so that e_k exceeds e_{sp} at a stated level of probability p_e. It should be noted here that the uncertainty associated with the event ($e_k > e_{sp}$) arises as a result of imperfect *knowledge* of the material supplied and the difficulties of obtaining sufficient sample data at the appropriate time. The probability p_e must be clearly distinguished from the probability p mentioned above.

The *design value* e_d of the strength of a material is obtained from the specified or characteristic strength as follows

$$e_d = \frac{e_{sp}}{\gamma_m} \tag{11.12}$$

where γ_m is a partial coefficient on strength.

11.3.5 Treatment of loads and other actions

The classification and modelling of loads and other actions were discussed in chapter 3. Most loads differ from other basic variables in that they vary significantly with time and are generally not amenable to effective control. There are some notable exceptions to both these generalisations.

Because of the time-varying nature of most loads, the problem of assessing the combined effect of a number of different loads acting on a structure has been seen so arise. This was discussed in chapter 10 in the context of reliability analysis. As might be expected a rather similar problem arises in treating combined loads within the framework of a deterministic level 1 code.

Characteristic values: The uncertainty in most permanent loads is small and for this reason it is customary to use the mean or nominal values of permanent loads in most design calculations. For the same reason it is appropriate that the characteristic value g_k of each permanent load G is taken as its mean value μ_G. μ_G may be considered to be the average permanent load taken over all nominally similar structures and obtained by using mean dimensions and mean densities.

For a time-varying load Q, the characteristic value q_k is normally defined as that value which has a prescribed probability p of not being exceeded within a given reference period. It is therefore the pth fractile of the extreme value distribution of the load corresponding to that reference period. Up to the present date (1982) few national loading committees have attempted to rationalise their specified loads along these lines, but progress is being made in this direction. Wind loading codes are perhaps the most advanced in this respect, e.g. [11.2]. The nominal loads specified in most loading codes vary rather widely in terms of their probability of exceedance.

Single time-varying loads: If a structure or structural component is subjected to only permanent loads G and one time-varying load Q, the load-combination problem does not arise. In this case, the values g_d and q_d to be used in the design or safety checking process (cf. equation (11.6)) are obtained from

$$g_d = \gamma_{fG} g_k \tag{11.13}$$

$$q_d = \gamma_{fQ} q_k \tag{11.14}$$

where γ_{fG} and γ_{fQ} are partial coefficients and g_k and q_k are characteristic values of the random variables G and Q, respectively.

For failure modes in which part of the permenent load acts in a stabilising or resisting sense and part in a de-stabilising or loading sense, different values of γ_{fG} should be used for the two components; $\gamma_{fG} \leqslant 1$ when the load is stabilising the structure and $\gamma_{fG} \geqslant 1$ when it is not.

Combinations of time-varying loads: When a structure has to resist a number of stochastically independent time-varying loads, it is clear that the probability of two or more loads exceeding their characteristic values simultaneously is small. If the total load effect in a member were to be determined from

$$S = c(\gamma_{fG1} g_{k1}, \ldots, \gamma_{fGm} g_{km}, \gamma_{fQ1} q_{k1}, \ldots, \gamma_{fQn} q_{kn}) \tag{11.15}$$

where

γ_{fG1} and g_{k1} are the values of γ_{fG} and q_k for the first of m permanent loads,

q_{k1} is the characteristic value of the first of n time-varying loads Q_i,

γ_{fQ1} is the partial coefficient associated with the load Q_1 *when this load is acting alone*, and

c is the load effect function, implying a linear or, where appropriate, a non-linear analysis of the structure under the action of the factored loads,

the resulting load effect S would be extremely conservative. For this reason it is necessary to introduce a set of reduction factors ψ_{0i} ($\psi_{0i} \leqslant 1$) to be applied to the time-varying loads Q_i to take account of the reduced probability of the design values of the loads being exceeded simultaneously. The total design load effect is therefore given by (cf. Turkstra's rule, p. 168),

$$S_d = c\,(\gamma_{fG1} g_{k1}, \ldots, \gamma_{fGm} g_{km}, \gamma_{fQ1} \psi_{01} q_{k1}, \ldots, \gamma_{fQn} \psi_{0n} q_{kn}) \tag{11.16}$$

In principle, if there are n time-varying loads, it is necessary to undertake n design checks (equation (11.6)) on the structure, using a separate set of ψ_0 factors for each check and with $\psi_{0jj} = 1$ for the jth check.

For the jth design check equation (11.16) may then be re-written as

$$S_d = c\,(g_{d1}, \ldots, g_{di}, \ldots, g_{dm}, q_{d1}, \ldots, q_{di}, \ldots, q_{dj}, \ldots, q_{dn})$$

$$= s\,(\bar{g}_d, \bar{q}_d, \bar{d}_{Sd}, \bar{c}) \tag{11.17}$$

where

$$g_{di} = \gamma_{fGi} g_{ki}$$

$$q_{di} = \gamma_{fQi} \psi_{0ji} q_{ki}$$

$$q_{dj} = \gamma_{fQj} \psi_{0jj} q_{kj} = \gamma_{fQj} q_{kj} \quad \text{and}$$

$$\bar{g}_d = (g_{d1}, \ldots, g_{dm}), \text{ etc.}$$

The need for a number of design checks using different sets of ψ_0 factors arises from the fact that throughout a structure the contribution of each separate load Q_i to the maximum load-effect in any member, varies considerably from member to member. For example, although snow loading may dominate the load effect in the roof beams of a multi-storey building, the same loads have only a small influence on the total load effects in the ground floor columns.

In practice, with detailed knowledge of the structure being designed or checked, it is often possible to reduce the number of safety checks significantly.

Equations (11.6) and (11.16) are the most general form of checking equations that are envisaged for use in level 1 codes. Some rather less general forms of checking equations have also been suggested [11.7]. In practical codes the design requirements may be made considerably simpler.

11.4 METHODS FOR THE EVALUATION OF PARTIAL COEFFICIENTS

Any reader who is unfamiliar with the theory of level 1 codes may be somewhat concerned by the apparent complexity of the safety checking rules set out in section 11.3 and by the apparent arbitrariness of some of the steps. Because of the inherently probabilistic nature of most structural safety problems, it is clear that safety checking procedures which are couched in deterministic terms will have some degree of arbitrariness. This cannot be avoided.

The design clauses given in level 1 codes should be interpreted as a set of *decision rules*, the outcome of which can be modified by changes to a set of *control parameters* - the partial coefficients. The process of selecting the set of partial coefficients to be used in a particular code should be seen as a process of optimization such that the outcome of all designs undertaken to the code is in some sense optimal. This should not be confused with the concept of optimizing individual structures. Whether or not a formal optimization is undertaken in practice, it is useful to think of the partial coefficient selection process in this way. It is then clear that the possibility exists for using any simplified set of design clauses together with a modified set of partial coefficients which on average will achieve the same degree of safety as the more complex set. The penalty to be paid for using the simplified design rules is some increase in materials usage.

In the remainder of this section various formal procedures for the determination of partial coefficients are discussed.

11.4.1 Relationship of partial coefficients to level 2 design point

It was shown in chapter 5 that for the reliability analysis of a particular structure, the level 2 method involves the mapping of the set of n basic random variables $\overline{X}$ to a set of independent standard normal variables $\overline{Z}$. This results in the mapping of the limit state failure surface given by

$$g(x_1, x_2, \ldots, x_n) = 0 \tag{11.18}$$

to a failure surface in standard normal space

$$f(z_1, z_2, \ldots, z_n) = 0 \tag{11.19}$$

The reliability index β is defined in Z space as the shortest distance from the origin to the failure surface and is given by (see (5.34))

$$\beta = \min_{\overline{z} \in \partial\omega} \left(\sum_{i=1}^{n} z_i^2\right)^{\frac{1}{2}} \tag{11.20}$$

The point on the failure surface which is closest to the origin is referred to as the design point (see figure 5.5) and has co-ordinates $(\beta\alpha_1, \beta\alpha_2, \ldots, \beta\alpha_n)$, where (see (5.35))

$$\alpha_i = -\left[\sum_{k=1}^{n} \frac{\partial f}{\partial z_k}(\beta\bar{\alpha}))^2\right]^{-\frac{1}{2}} \frac{\partial f}{\partial z_i}(\beta\bar{\alpha}) \quad , \quad i = 1, 2, \ldots, n \tag{11.21}$$

with

$$f(z_1^\star, z_2^\star, \ldots, z_n^\star) = 0 \tag{11.22}$$

and $\bar{z}^\star = \beta\bar{\alpha}$.

By using the inverse mapping

$$x_i^\star = F_{X_i}^{-1}(\Phi(z_i^\star)) \quad , \quad i = 1, 2, \ldots, n \tag{11.23}$$

we obtain the set of values $\bar{x}^\star$ for the original basic variables $\bar{X}$ corresponding to the design point $\bar{z}^\star$. If the variables $\bar{X}$ are all normally distributed, then the set of values $\bar{x}^\star$ are the values of the variables at which failure is most likely to occur (if this event were to happen), i.e.

$$\max_{\bar{x} \in \omega_f} f_{X_i}(x_i) = f_{X_i}(x_i^\star) \quad , \quad i = 1, 2, \ldots, n \tag{11.24}$$

where ω_f is the failure region.

If $\bar{X}$ are non-normal then equation (11.24) is only approximate.

It can now be seen that if the values $\bar{x}^\star$ were to be used as the design values $\bar{x}_d$ in a deterministic level 1 design calculation, the resulting structure would have a reliability index β and a reliability $\mathcal{R} = 1 - \Phi(-\beta)$. Thus, if $\mathcal{R}$ is an acceptable reliability for the structure, a satisfactory set of partial coefficients is given by

$$\gamma_i = \frac{x_{sp_i}}{x_{d_i}} = \frac{x_{sp_i}}{x_i^\star} = \frac{x_{sp_i}}{F_{X_i}^{-1}(\Phi(z_i^\star))} \tag{11.25}$$

where x_{sp_i} is the specified value of the *resistance variable* X_i, and by

$$\gamma_j = \frac{x_{d_j}}{x_{sp_j}} = \frac{x_j^\star}{x_{sp_j}} = \frac{F_{X_j}^{-1}(\Phi(z_j^\star))}{x_{sp_j}} \tag{11.26}$$

where x_{sp_j} is the specified value of the *loading variable* X_j.

Example 11.1. If X_j is a normally distributed loading variable, then

$$\gamma_j = \frac{F_{X_j}^{-1}(\Phi(z_j^\star))}{x_{sp_j}} = \frac{\mu_{X_j} + \alpha_j\beta\sigma_{X_j}}{x_{sp_j}} \tag{11.27}$$

: Assuming that the parameters μ_{X_j} and σ_{X_j} of the variable X_j are known or can be esti-
: mated, that x_{sp_j} is given and that the reliability index β is specified, the evaluation of the
: partial coefficient γ_j requires only a knowledge of the sensitivity factor α_j.

: **Example 11.2.** If X_i is a log-normally distributed resistance variable, then

$$\gamma_i = \frac{x_{sp_i}}{F_{X_i}^{-1}(\Phi(z_i^*))} = \frac{x_{sp_i}}{\mu_{X_i}\exp(-\frac{1}{2}\ell n(V_i^2+1) + \alpha_i\beta(\ell n(V_i^2+1))^{\frac{1}{2}})}$$

$$\simeq \frac{x_{sp_i}}{\mu_{X_i}\exp(\alpha_i\beta V_i)} \tag{11.28}$$

: where

: μ_{X_i} is the mean of X_i, and

: V_i is the coefficient of variation of X_i.

: Again, if the parameters μ_{X_i} and V_i are known, and x_{sp_i} and β are given, then γ_j can be
: evaluated from a knowledge of the sensitivity factor α_i.

Equations (11.27) and (11.28) and similar relationships for other types of probability distribution are only of direct use, when the values $\bar{\alpha}$ are known. In general, the value of α_i depends not only on the parameters of the random variable X_i, but on the values of the parameters of the other random variables, on the value of β and on the nature of the limit state function.

For a particular structure and failure mode, the sensitivity factors $\bar{\alpha}$ may be evaluated from equation (11.21). However, the use of this equation implies a reliability analysis of the structure and if this is to be undertaken there is little point in following it with a level 1 safety check. Furthermore, this approach leads to a partial coefficient on every basic variable, which is too many for practical use in design.

A procedure is therefore required for the determination of a limited number of partial coefficients or additive safety elements ($\ll$ n, where n is the number of basic variables) which will be applicable over a range of different failure modes and for a range of different structural types covered by a code of practice. Such a procedure is discussed in section 11.4.3. Before this, we shall consider an approximate direct method for the evaluation of partial coefficients.

11.4.2 Approximate direct method for the evaluation of partial coefficients

The difficulty with the approach suggested above was seen to lie in the evaluation of suitable sensitivity factors $\bar{\alpha}$. Experience shows that over fairly large ranges of design parameters the individual factors α_i often do not change dramatically. Furthermore, because

$$\sum_{i=1}^{n} \alpha_i^2 = 1$$

it is always possible to choose a conservative set of sensitivity factors for use with equation

(11.26). $\alpha_1 = \pm \alpha_2 = \ldots = \pm \alpha_n = \pm 1$ is such a set, when the sign of the factor is taken as positive for loading variables and negative for resisting variables; although in most practical cases this would be too conservative.

Assuming that the limit state function may be split into a resistance term R and a load effect term S, as in equation (11.6), it has been proposed [11.5] that the sensitivity factors should be expressed as

$$\alpha_{R,i} \triangleq \tilde{\alpha}_R \hat{\alpha}_{R,i} \tag{11.29}$$

$$\alpha_{S,i} \triangleq \tilde{\alpha}_S \hat{\alpha}_{S,i} \tag{11.30}$$

where

$\alpha_{R,i}$ is the sensitivity factor for the ith resistance variable,

$\alpha_{S,i}$ is the sensitivity factor for the ith loading variable,

$\tilde{\alpha}_R$ and $\tilde{\alpha}_S$ are estimates of the sensitivity factor for the composite variables R and S in the limit state function $R - S = 0$,

$\hat{\alpha}_{R,i}$ is a factor which depends on the relative importance of the ith resistance variable, and

$\hat{\alpha}_{S,i}$ is a factor which depends on the relative importance of the ith loading variable.

Assuming that the true values $\bar{\alpha}$ are known (i.e. from a level 2 analysis) and the variables $\overline{X}$ are ranked (taking due account of sign) so that

$$\alpha_{R,1} < \alpha_{R,2} < \ldots < \alpha_{R,n_R} \quad , \quad -1 \leqslant \alpha_{R,i} \leqslant 0 \tag{11.31}$$

$$\alpha_{S,1} > \alpha_{S,2} > \ldots > \alpha_{S,n_S} \quad , \quad 0 \leqslant \alpha_{S,i} \leqslant 1 \tag{11.32}$$

where $n_R + n_S = n$, the total number of basic variables, the quantities R_1 and S_1 may be termed the leading resisting and loading variables, respectively.

For a wide range of structural members, the following empirically-based values can be shown to be satisfactory

$$\tilde{\alpha}_R = 0.8 \qquad \tilde{\alpha}_S = 0.7 \tag{11.33}$$

$$\hat{\alpha}_{R,i} = \sqrt{i} - \sqrt{i-1} \quad , \quad i = 1, 2, \ldots, n_R \tag{11.34}$$

$$\hat{\alpha}_{S,i} = \sqrt{i} - \sqrt{i-1} \quad , \quad i = 1, 2, \ldots, n_S \tag{11.35}$$

Hence, for the loading variables R_1 and S_1, $\hat{\alpha}_{R,1} = \hat{\alpha}_{S,1} = 1$ giving $\alpha_{R,1} = -0.8$ and $\alpha_{S,1} = 0.7$. This approach is viable only if the designer has prior knowledge of the relative importance (sensitivity ranking) of the various variables. This information can be gained by experience and by the occasional level 2 analysis.

Having estimated the sensitivity factors $\bar{\alpha}$ from equations (11.29) to (11.35), the partial coefficients γ_i and γ_j, or the design values of the variables x_i^* and x_j^*, may be obtained directly from equations (11.25) and (11.26). This process is illustrated in the following simple example.

Example 11.3. The encastré steel beam shown in figure 11.3 is to be designed against plastic collapse to resist a uniformly distributed superimposed load Q and a permanent load G. Q, G, the yield stress of the steel E_y, and the model uncertainty X_m affecting the plastic moment of resistance of the section are assumed to be normally distributed random variables, with the parameters given in table 11.1. The yield stresses at the plastic hinge positions A, B and C are assumed to be the same and the geometrical variables are assumed to have no uncertainty. It is desired to evaluate the partial coefficients γ_Q, γ_G, γ_{E_y} and γ_{X_m} for a reliability index $\beta = 4$, and to determine the required plastic modulus z_p.

By consideration of the mean values and coefficients of variation of the variables and the nature of the limit state function it may be assumed that

$$\alpha_Q > \alpha_{X_m} > \alpha_{E_y} > \alpha_G$$

Thus,

$$\alpha_Q = \tilde{\alpha}_S \hat{\alpha}_{S,1} = 0.7 \times 1.0 = 0.7$$

$$\alpha_{E_y} = \tilde{\alpha}_R \hat{\alpha}_{R,1} = -0.8 \times 1.0 = -0.8$$

$$\alpha_{X_m} = \tilde{\alpha}_R \hat{\alpha}_{R,2} = -0.8 \times (\sqrt{2} - 1) = -0.331$$

$$\alpha_G = \tilde{\alpha}_S \hat{\alpha}_{S,2} = 0.7 \times (\sqrt{2} - 1) = 0.290$$

Variable	μ_X	σ_X	V_X	x_{sp}
Q kN/m	40.1	6.015	15%	50.0
G kN/m	30.0	1.5	5%	30.0
E_y N/mm²	293.6	23.49	8%	255.0
X_m	1.0	0.06	6%	1.0

Table 11.1

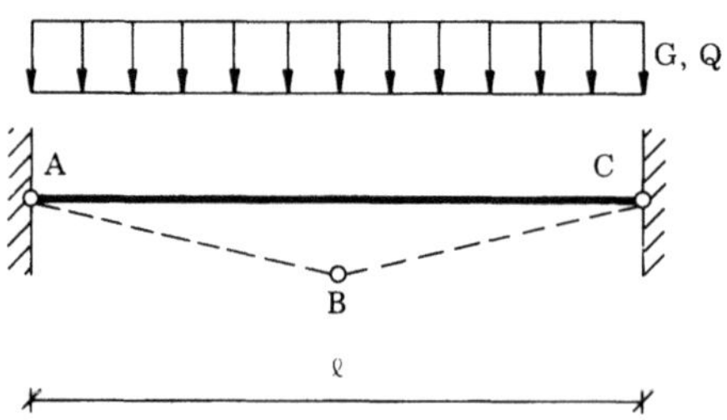

Figure 11.3

: and the design values x^* are given by

$$q^* = \mu_Q + \alpha_Q \beta \sigma_Q = 56.94 \text{ kN/m}$$

$$e_y^* = \mu_{E_y} + \alpha_{E_y} \beta \sigma_{E_y} = 218.4 \text{ N/mm}^2$$

$$x_m^* = \mu_{X_m} + \alpha_{X_m} \beta \sigma_{X_m} = 0.921$$

$$g^* = \mu_G + \alpha_G \beta \sigma_G = 31.74 \text{ kN/m}$$

: These values and the partial coefficients found from equations (11.25) and (11.26) are listed in table 11.2.

Variable	x_{sp}	x^*	γ
Q kN/m	50.0	56.9	1.13
G kN/m	30.0	31.7	1.06
E_y N/mm²	255.0	218.4	1.17
X_m	1.0	0.92	1.09

Table 11.2

: By application of virtual work, the required plastic modulus z_p may now be determined from

$$(\gamma_G g_{sp} + \gamma_Q q_{sp}) \frac{\ell^2}{4} = 4(\frac{e_{sp}}{\gamma_{E_y}} z_p) \frac{10^{-6}}{\gamma_{X_m}} \qquad (11.36)$$

: Substituting the appropriate values from table 11.2 gives $z_p = 6.89 \times 10^5 \text{ mm}^3$.

: Finally, it is of interest to use the level 2 method to determine the reliability of this structure when the plastic modulus has the value found by the above method. The failure function can be written as

$$M = 16\, z_p E_y X_m - \ell^2 (G + Q) \times 10^6 = 0 \qquad (11.37)$$

: Using the methods of chapter 5 and the parameters from table 11.1 gives a reliability index $\beta = 4.45$. This is larger than the originally selected value of 4.0 showing that the approximate method of determining partial coefficients is safe, at least for the structure and set of variables examined.

It must be stressed that great care must be taken when using the spproximate method for the evaluation of partial coefficients if the relative magnitudes of the sensitivity factors $\bar{\alpha}$ are not known. Care must also be taken when there is appreciable statistical uncertainty in the parameters of the probability distributions of the basic variables because of lack of data. Methods of including statistical uncertainty were introduced in section 7 of chapter 3. See also [11.11].

11.4.3 General method for the evaluation of partial coefficients

Practical codes should have the smallest number of partial coefficients that is consistent with reasonably uniform standards of reliability; moreover, the same partial coefficients should be applicable to a wide range of structural components. This means that they must be applicable over a range of sensitivity factors without being unsafe or unduly conservative. A suitable general method for the evaluation of such a set of partial coefficients is now presented.

The first stage of this process is to decide upon an appropriate standard of reliability or *target failure probability* for the structures (or more generally, structural components, e.g. beams, columns, slabs) that will be designed using the new code. This is also a pre-requisite for the procedure described in section 11.4.2. The choice is generally made by a process of probabilistic calibration to an existing code. e.g. see [11.6].

Studies of the reliability of structural components designed to traditional codes typically show very wide ranges of reliability. An appropriate choice for the target failure probability P_{ft} for a new code is the *weighted average* of the failure probabilities exhibited by components designed to existing codes, provided that the least reliable component exhibited satisfactory performance in actual service. The latter is not always easy to verify because existing codes may not have been in use for a sufficiently long period of time and structures may have been subjected to only a fraction of their design loads. The *weighting factors* ω_i should be selected to represent the previous frequency of usage of each structural component included in the calibration and should be such that $\sum_i \omega_i = 1.0$.

Use of the weighted average failure probability rather than, say, the weighted average reliability index means that the target failure probability P_{ft} tends to be governed by the less reliable components in existing codes. This assumes a measure of economy in the new code, but care has to be taken that these reliabilities are not too low.

A more direct approach to the choice of target failure probabilities has been recommended by the Nordic Committee on Building Regulations (NKB) [11.10]. In this, the target failure probability depends on the consequences of failure and on the nature of the failure mode, as shown in table 11.3.

Failure consequences	Failure type I	Failure type II	Failure type III
Not serious	10^{-3} / 3.09	10^{-4} / 3.71	10^{-5} / 4.26
Serious	10^{-4} / 3.71	10^{-5} / 4.26	10^{-6} / 4.75
Very serious	10^{-5} / 4.26	10^{-6} / 4.75	10^{-7} / 5.20

Table 11.3. Target failure probabilities and corresponding reliability indices [11.10].

The target failure probabilities given in table 11.3 are for a reference period of 1 year, but should be treated as operational or notional probabilities and not as relative frequencies. The failure types are defined as

I ductile failure with reserve strength capacity resulting from strain hardening

II ductile failure with no reserve capacity

III brittle failure and instability

Having chosen a target failure probability, the problem of selecting a set of partial coefficients $\bar{\gamma}$ for a code, or part of a code, may now be reduced to the application of the following simple principle. Choose the set of partial coefficients $\bar{\gamma}$, so as to minimise the quantity S given by

$$S = \sum_{i=1}^{m} \omega_i \Delta(P_{fi}(\bar{\gamma}), P_{ft}) \quad (11.38)$$

Subject to the constraint

$$\sum_{i=1}^{m} \omega_i P_{fi}(\bar{\gamma}) = P_{ft} \quad \text{with} \quad \sum_{i=1}^{m} \omega_i = 1.0 \quad (11.39)$$

and where

$\Delta(P_{fi}(\bar{\gamma}), P_{ft})$ is an agreed function of the quantities $P_{fi}(\bar{\gamma})$ and P_{ft},

$P_{fi}(\bar{\gamma})$ is the failure probability of the ith structural component designed using the set of partial coefficients $\bar{\gamma}$,

P_{ft} is the target failure probability,

$\bar{\omega} = (\omega_1, \ldots, \omega_m)$ is a set of weighting factors indicating the relative importance of each of the m structural components included in the partial factor evaluation.

In general terms, the aim of this approach is to minimise the deviations of the probabilities P_{fi} from the target probability of failure P_{ft}, whilst maintaining the average probability of failure at the target level. Experience has shown that the values of the partial coefficients are generally very insensitive to the form of the objective function used (equation (11.38)). Suitable functions are:

$$S_1 = \sum_{i=1}^{m} \omega_i (\log P_{fi}(\bar{\gamma}) - \log P_{ft})^2 \quad (11.40)$$

and

$$S_2 = \sum_{i=1}^{m} \omega_i (-\Phi^{-1}(P_{fi}(\bar{\gamma})) + \Phi^{-1}(P_{ft}))^2 = \sum_{i=1}^{m} \omega_i (\beta_i(\bar{\gamma}) - \beta_t)^2 \quad (11.41)$$

where β is the reliability index.

Clearly, many other possibilities exist. Obtaining the solution to equations (11.38) and (11.39) is a problem of constrained minimisation for which a number of standard techniques and computer programs are available. Nevertheless, the total amount of computational effect is considerable because all the probabilities P_{fi} need to be re-evaluated for each adjustment to the partial coefficients $\bar{\gamma}$.

The code writer is free to choose as many partial coefficients or additive safety elements as is considered appropriate for a given code. A practical number is generally considerably less than the number of basic random variables. A reduction in the number of partial coefficients can be achieved by constraining the unwanted coefficients to be unity. Provided equation (11.39) is satisfied, the effect of these additional constraints is to increase the deviations from the target failure probability P_{ft} and to increase the average amount of material used when designing to the code. The penalty to be paid for increased simplicity in the code safety format is therefore some increase in the initial cost of construction.

When applying this procedure over a number of codes for different construction materials, e.g. steel and concrete, a further constraint that should be considered is to make the partial coefficients on loads and other actions the same in each code, irrespective of construction material. Such an approach has many practical advantages.

Finally, a note of caution. In chapter 5 it was mentioned that the reliability index as defined by equation (5.9) is not invariant with regard to the choice of failure function. A similar problem of lack of invariance arises when the partial coefficients used in a code are not directly associated with their corresponding sources of uncertainty. This occurs when the number of partial coefficients is constrained to some small number. In such cases, the partial coefficients should be used only with the precise form of the design equations (failure functions) for which they were derived.

11.5 AN EXAMPLE OF PROBABILISTIC CODE CALIBRATION

The general method for the evaluation of partial coefficients which was described in the previous section involves a considerable amount of effort and computation and is not easily illustrated by a simple example. For this reason, some results that were obtained during the probabilistic calibration [11.6] of the U.K. Steel Bridge Code BS 5400: Part 3 [11.3] are included here as an illustration of the method.

11.5.1 Aims of calibration

BS 5400: Part 3 is a level 1 code in which the degree of structural reliability is contolled by a number of partial coefficients (partial factors). The code replaces an earlier British Standard, BS 153 [11.1] and was developed mainly for the purposes of incorporating technical improvements in many of the design clauses; but at the same time the opportunity was taken to rationalise the safety provisions and to change from a permissible stress to a limit state approach.

In evaluating the partial coefficients, the agreed policy was to achieve the same *average* reliability for components designed to the new code as the *average* inherent in designs to the previous code BS 153, but at the same time to reduce the scatter in the reliability of the various components. An obvious limitation of this work is that it was restricted to a study of structural components rather than structural systems.

A flow-chart showing the various steps in the calculation procedure is given in figure 11.4.

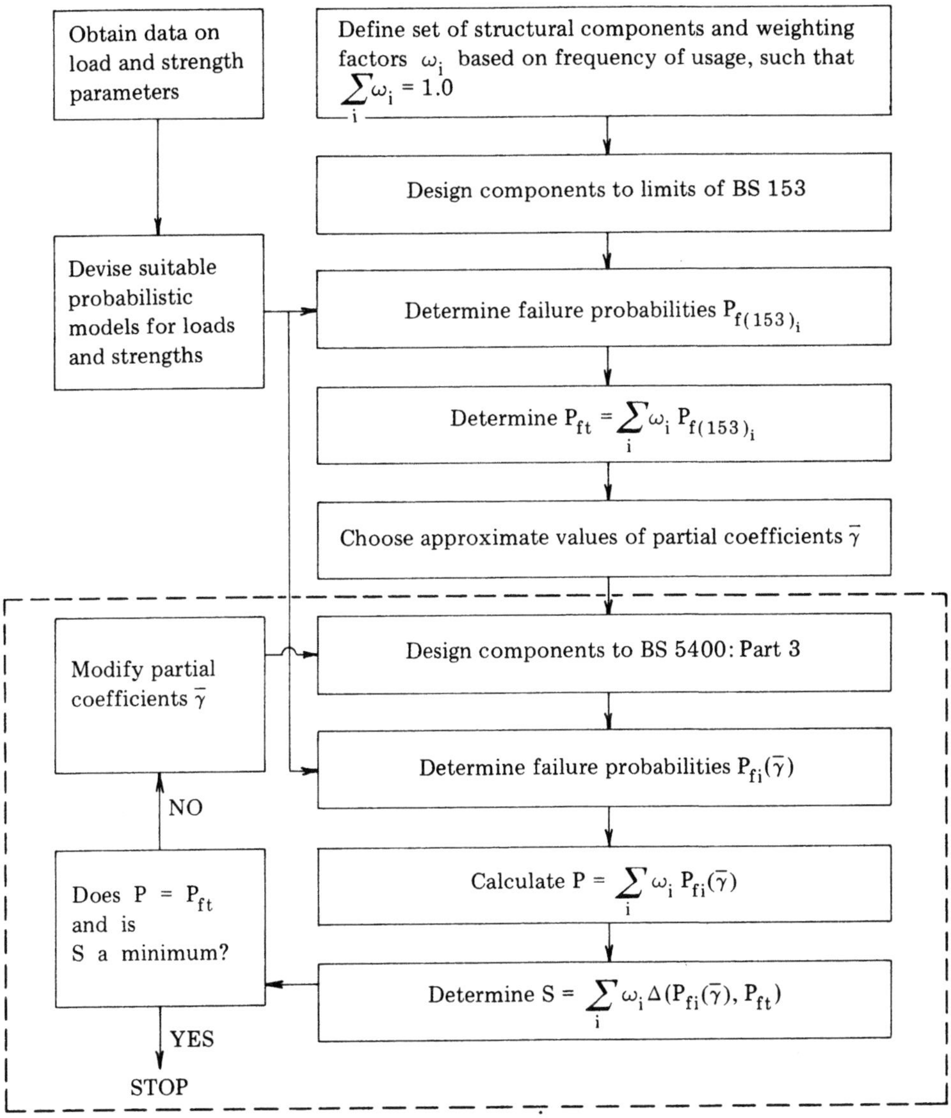

Figure 11.4. Probabilistic calibration of BS 5400: Part 3 to BS 153.

11.5.2 Results of calibration

Figure 11.5 shows the scatter in the computed failure probabilities for the major structural components designed to the limits of BS 153 which were included in the calibration calculations. The failure probabilities exhibit very wide scatter varying over many orders of magnitude. In addition, there are significant differences in the average reliability of different types of component. Neither of these facts is surprising since the code was originally based on deterministic concepts with no regard for the relative magnitude of the various uncertainties. It should be noted that the modelling did not allow for the possibility of gross errors in design or construction and for this reason the probabilities should be interpreted as a measure of relative safety and not as failure frequencies.

The target failure probability P_{ft} for the new code BS 5400: Part 3 was determined as the weighted average of the failure probabilities for components designed to BS 153 and was 0.63×10^{-6}. In calculating this value, stiffened compression flanges and unwelded plate panels were excluded, the former because they had not been shown to behave satisfactorily in service and the latter because the data on model uncertainty were considered inadequate.

The partial coefficients for the new code were determined for use with checking equations of the form

$$1/\gamma_{m2}\ \text{function}[f_y/\gamma_{m1},\ \text{other parameters}] - \text{effects of } [\gamma_{fG1}G_1,\ \gamma_{fG2}G_2,\ \gamma_{fQ}Q] > 0 \tag{11.42}$$

where

f_y is the yield stress of the steel,

G_1 is structural self weight,

G_2 is superimposed permanent load,

Q is traffic loading,

γ_{m1} is a partial coefficient on yield stress which applies throughout the code,

γ_{m2} is a partial coefficient on the computed resistance which varies with type of component, and

γ_{fG1}, γ_{fG2} and γ_{fQ} are partial coefficients on loads.

Values of the partial coefficients obtained by minimising the quantity S defined by equation (11.38), subject to the constraint given by equation (11.39), are listed in column 1 of table 11.4. The other columns in this table show the values of partial coefficients γ_{m2} when other constraints are introduced. For example, column 3 shows the effect of setting $\gamma_{m1} = 1.0$, $\gamma_{fQ} = 1.5$ and $\gamma_{fG} = 1.13$ (given here as a weighted average of γ_{fG1} and γ_{fG2}), and thus effectively eliminating γ_{m1} from the code.

The quantity S is given in the penultimate row of table 11.4 for each of the sets of partial coefficients calculated and can be seen to increase as additional constraints are introduced. The quantity $\sum_i \omega_i \theta_i$ given in the last row of the table is the ratio of the amount of steel used when

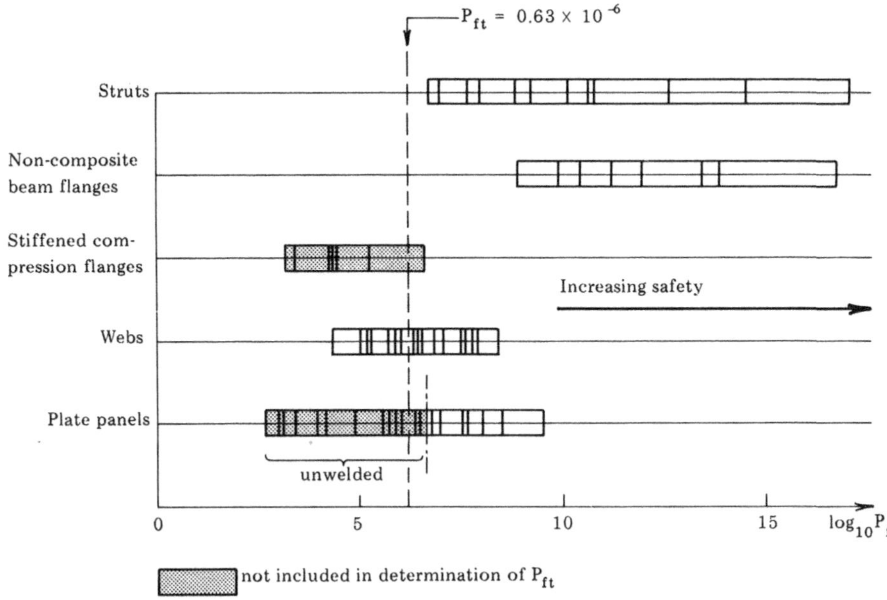

Figure 11.5. Failure probabilities for components designed to BS 153 and early flange rules (from |11.6|).

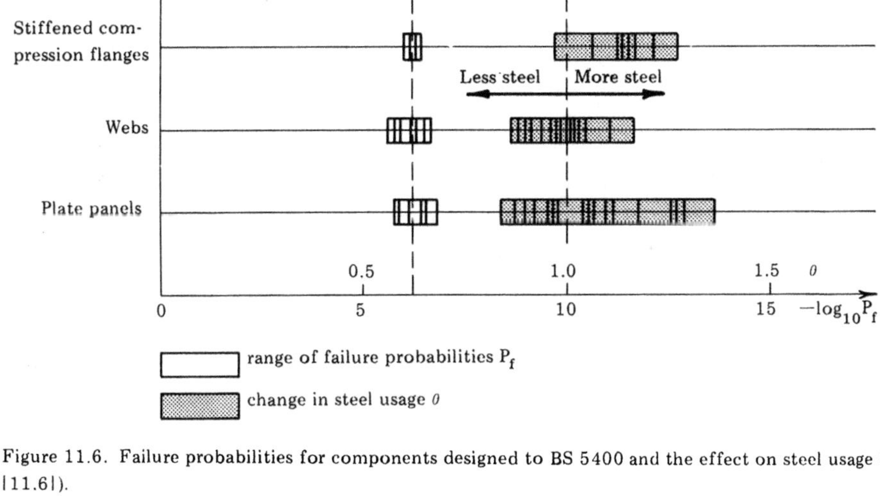

Figure 11.6. Failure probabilities for components designed to BS 5400 and the effect on steel usage (from |11.6|).

	1	2	3	4	5	6	7
	Fully optimised coeffs.	Increasing constraints →					
γ_{fG}	1.16	1.13	1.13	1.13	1.13	1.13	1.13
γ_{fQ}	1.47	1.50	1.50	1.50	1.50	1.50	1.50
γ_{m1}	1.08	1.08	1.0	1.0	1.0	1.0	1.0
γ_{m2} for:							
Struts	0.98	0.98	1.03	1.05	1.0	1.3	1.35
Beam flanges	1.09	1.08	1.17	1.15	1.2		
Stiffened compression flanges	1.27	1.28	1.37	1.35	1.4		
Webs	1.25	1.25	1.34	1.35	1.3		
Plate panels	1.08	1.08	1.14	1.15	1.1		
Ties	1.09	1.09	1.18	1.20	1.2		
$\sum_i \omega_i p_{fi} [\times 10^{-6}]$	.632	.632	.632	.658	1.146	1.072	0.288
S	.073	.086	.142	.225	.282	5.95	8.80
$\sum_i \omega_i \theta_i$	.936	.938	.939	.942	.933	1.00	1.04

← Calculated coefficients | Rounded or arbitrary coefficients →

Table 11.4. Partial coefficients for various degrees of constraint (from [11.6]).

designing with the new code (with the partial coefficients given) to the amount used when designing with original code BS 153. Use of any of the sets of partial coefficients in columns 1-4 would therefore give a saving of approximately 6% in steel consumption compared with BS 153. Further constraints on the number of partial coefficients used reduces this saving. In practice, the balance between the simplicity of the safety format and the savings in material must be decided by the code-writing committee.

No code calibration study should be considered to be complete until the effects of the change in safety format on the design of all components within the scope of the code have been examined. It is important to know the range of failure probabilities for each type of component when using the proposed set of partial coefficients. It is also of interest to know the changes in the quantities of materials that will be used compared with earlier codes. The latter is perhaps the most tangible measure of change in safety levels. This is illustrated in figure 11.6.

BIBLIOGRAPHY

[11.1] British Standards Institution: *Specification for Steel Girder Bridges.* BS 153, 1958.

[11.2] British Standards Institution: *Basic Data for the Design of Buildings. Wind Loads.* C.P. 3: Chapter V, Part 2, 1972.

[11.3] British Standards Institution: *Steel, Concrete and Composite Bridges.* Part 3: Code of Practice for Design of Steel Bridges. BS 5400, Part 3, 1982.

[11.4] CIRIA: *Rationalisation of Safety and Serviceability Factors in Structural Codes.* Construction Industry Research and Information Association, Report No. 63, 1977.

[11.5] Euro-Code No. 1: *Common Unified Rules for Different Types of Constructions and Material.* Appendix II: Guidelines for Determining Partial Coefficients. Draft, February, 1981.

[11.6] Flint, A. R., Smith, B. W., Baker, M. J. and Manners, W.: *The Derivation of Safety Factors for Design of Highway Bridges.* Proc. Conf. on The New Code for the Design of Steel Bridges, Cardiff, March 1980. Granada Publishing, 1981.

[11.7] Joint Committee on Structural Safety, CEB - CECM - CIB - FIP - IABSE - IASS - RILEM: *International System of Unified Standard Codes for Structures.* Volume I: Common Unified Rules for Different Types of Construction and Material. CEB/FIP, 1978.

[11.8] Joint Committee on Structural Safety, CEB - CECM - CIB - FIP - IABSE - IASS - RILEM: *General Principles on Reliability for Structural Design.* International Association for Bridge and Structural Engineering, 1981.

[11.9] Lind, N. C.: *Reliability-Based Structural Codes - Optimisation Theory.* International Research Seminar on Safety of Structures under Dynamic Loading. Vol. 1. Trondheim, June, 1977.

[11.10] Nordic Committee on Building Regulations: *Recommendation for Loading and Safety Regulations for Structural Design*. NKB-Report No. 36, Nov. 1978.

[11.11] Rackwitz, R.: *Implementation of Probabilistic Safety Concepts in Design and Organisational Codes*. Proceedings of the 3rd International Conference on Structural Safety and Reliability, Trondheim, 1981. Elsevier, 1981.

[11.12] Parimi, S. R. and Lind, N. C.: *Limit State Basis for Cold Formed Steel Design*. Proc. ASCE, J. Struct. Div., March 1976, Vol. 102, No. ST3.

Chapter 12

APPLICATIONS TO FIXED OFFSHORE STRUCTURES

12.1 INTRODUCTION

In the previous chapter, structural reliability theory was found to be a valuable tool in the development of rational safety formats for design codes and for the evaluation of partial coefficients. This is a major area of application. However, the main purpose of codes is to ensure that structures are designed and built in such a way that they can safety resist the (generally uncertain) loads to which they will be subjected; and it is therefore logical that reliability theory should be used directly as part of the design/analysis process, when appropriate. The main potential for direct application is with structures having large failure consequences or where the use gives immediate savings in construction costs. Other areas of application are in the assessment of existing structures and in research.

The examples given in earlier chapters are mainly of a simple nature and are included to illustrate the various calculation methods. The purpose of the present chapter is to discuss more complex structures where the direct application of the theory could be used in the process of safety assessment. Offshore structures have been chosen because of their economic importance and because they are the subject of much other research. Consideration will be restricted, however, to steel jacket structures, on which a number of detailed reliability studies have been carried out, e.g. [12.2], [12.26]. Nevertheless, many of the principles are equally applicable to other types of installation.

In the following, attention is focussed on the reliability of offshore structures under the action of extreme environmental loads (wind, waves and currents). Various aspects of the modelling of jacket structures are discussed, but *not* the broader features of offshore safety such as blow-outs, fires, explosions, collisions and other hazards. For information on these topics, see, for example, [12.15], [12.27].

12.2 MODELLING THE RESPONSE OF JACKET STRUCTURES FOR RELIABILITY ANALYSIS

Jacket structures are currently by far the commonest type of fixed offshore platform and are likely to continue to be used for oil and gas production in moderate water depths (up to 200

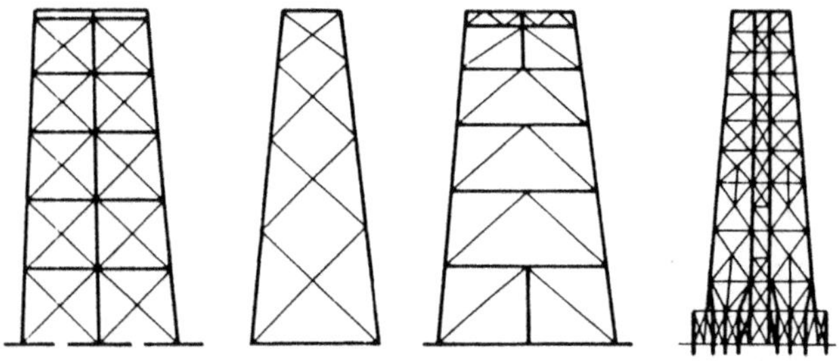

Figure 12.1. Elevations of various jacket structures [12.xx]

m) for the foreseeable future; although in greater depths, structures such as tension leg platforms (TLP's) may be found to be an economic proposition. Existing jacket structures range in height from a few tens of metres, as found in shallow coastal waters, to over 300 m. They also vary appreciably with regard to the number of main legs and the arrangement of bracing, with consequent effect on the degree of structural redundancy and the amount of local damage that can be sustained prior to collapse. Elevations of a number of different jacket structures (drawn to different scales) are shown in figure 12.1.

Structures such as that shown in figure 12.1 (d), the Cognac platform in the Gulf of Mexico, are immensely complex, employing thousands of tubular members and welded connections in the main jacket alone. For reasons of size, the actual response of such structures to wind and wave loads will be subject to considerable uncertainty, but most of the principal sources of uncertainty are clearly identifiable.

A major difference between the early shallow-water jackets and more recent offshore structures is in their response to wave loading. The smaller structures tend to be relatively stiff and to carry smaller deck loads and are thus not subject to significant dynamic response. In addition, they tend to be situated in more sheltered environments. Modern deep-water structures, on the other hand, are likely to experience appreciable dynamic motions and the reliability of such structures cannot be determined without a full dynamic analysis.

The reliability analysis of a large offshore jacket involves the same steps that must be undertaken in a conventional stochastic dynamic analysis, but in addition the main structural and loading variables and the parameters of the wave spectra are treated as random quantities.

The sources of uncertainty affecting the behaviour of dynamically sensitive jacket structures can be grouped under four main headings

- Those affecting the loading:
  - extreme wind speed
  - extreme current speed
  - the spectral form of the extreme sea-state
  - the extent of marine growth
  - hydrodynamic forces, given the water particle velocities and accelerations
  - permanent and semi-permanent deck loads
  - variable deck loads
- Those affecting the structural response:
  - the effect of uncertain soil properties on natural frequency
  - the effect of variable deck loads on natural frequency
  - structural and hydrodynamic damping
  - the peak response in a given reference period given the root mean square (r.m.s.) response
- Those affecting component strength:
  - material properties
  - geometrical imperfections
  - model uncertainties in predicted component strength
- Those affection system behaviour:
  - ductility
  - post-buckling strength of tubular members and other components

The purpose of a structural reliability analysis is to incorporate all these sources of uncertainty within a single set of calculations which can then be used to predict the relative likelihood of each of a large number of possible failure events. From this information and with knowledge of the consequences of the various failures, decisions can be made about the overall adequacy of the structure and the strength of individual components.

To be able to undertake these calculations it is first necessary to develop appropriate probabilistic models for the basic load and resistance variables, as discussed in chapter 3. In addition, further deterministic (or stochastic) models are required to relate the overall response of the structure to the applied loading, and to define the behaviour of the individual structural components in terms of the overall response. The word *model* is used here to emphasise the fact that the *actual behaviour* of all structures is extremely complex and that all calculation procedures involve a number of idealisations. For example, a typical jacket structure has a large (strictly infinite) number of degrees of freedom for dynamic response, but with careful modelling all the significant contributions to the response can be obtained by using a relatively small number of lumped masses.

Figure 12.2 shows the main models and calculation steps in the level 2 reliability analysis of a jacket structure using spectral analysis. The process illustrated is for the determination of the reliability of component members of the structure, conditional on no other members having failed, and where failure is of the first-passage type. For multi-element redundant

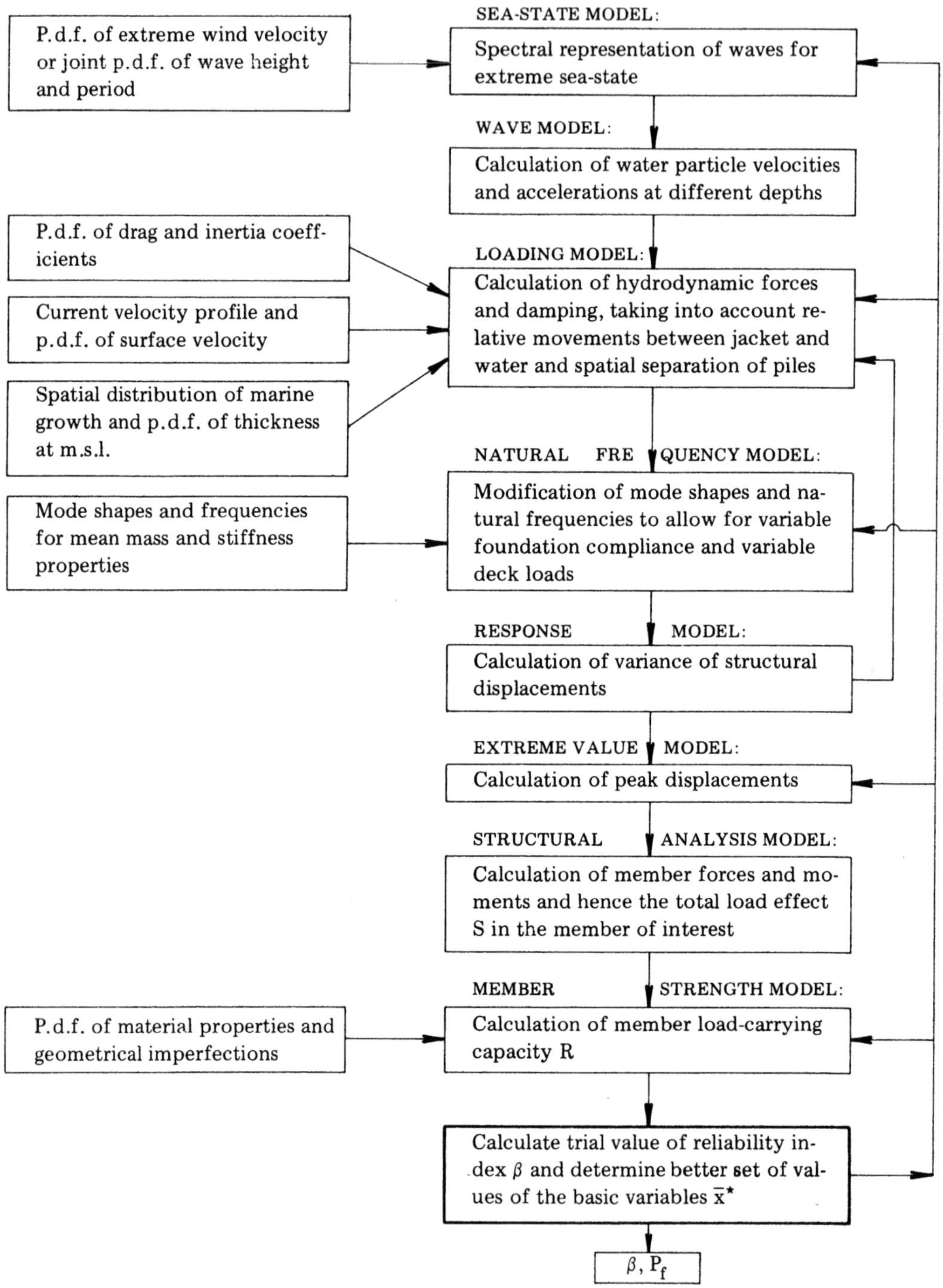

Figure 12.2. Models and calculation steps in the level 2 reliability analysis of a jacket structure using spectral analysis.

structures this is only the first stage of determining the system reliability, but the results can be used with the methods of chapters 7 and 8 to calculate upper bounds on the system failure probability. Accurate lower bounds can be found only by examining the sequences of member failures which result in complete collapse. This is the subject of active research and is beyond the scope of the present book. It should be noted, however, that the failures of different structures members are in general strongly correlated because of common loading variables. This has the effect of reducing the separation between the bounds. Moreover, for structures which are dynamically sensitive, the change in fundamental natural frequency following the »brittle» failure of just one major component (e.g. by fatigue or fracture) is likely to lead to a significant increase in dynamic response followed by rapid failure of other members within a short period of time.

In the following sub-sections a brief description is given of various loading and response models for dynamically sensitive offshore structures that can be used within a level 2 reliability analysis of such a structure.

12.2.1 Sea-state model

Waves are generated by wind blowing over the surface of the sea and are the major source of loading for most offshore structures. At any fixed position in the open sea, the level of the water surface varies randomly due to the passing waves and may be modelled as a stochastic process $X(t)$, where the index t denotes time.

As mentioned in chapter 9, the properties of $X(t)$ which are of greatest importance are the mean value function $\mu_X(t)$ defined by (see page 147)

$$\mu_X(t) = E[X(t)] = \int_{-\infty}^{\infty} x f_X(x; t) dx \tag{12.1}$$

and the autocorrelation function $R_{XX}(t_1, t_2)$ defined by

$$R_{XX}(t_1, t_2) = E[X(t_1)X(t_2)] = \int_{-\infty}^{\infty} x_1 x_2 f_{\{X\}}(x_1, x_2; t_1, t_2) dx_1 dx_2 \tag{12.2}$$

As previously discussed, when the first and second order distributions of $X(t)$ are independent of absolute time, the process is said to be weakly stationary. In this case, the autocorrelation function is dependent only on the time separation $\tau = t_1 - t_2$ and may be written as $R_{XX}(\tau)$.

It is clear that the level of the water surface at any position in the open sea is *not* a stationary process, because of changing meteorological conditions which are to some extent both seasonal and cyclic. However, for reasonably short periods of time (3-6 hours) the assumption of stationarity is not unreasonable. The reason for this can be seen by examining the frequency components of horizontal wind speed as shown in figure 12.3. In temperate latitudes major meteorological disturbances occur with a typical frequency of about 1 every 4 days and the wave conditions that are generated build up and die away at the same rate, with the worst conditions lasting about 6 hours. The process of wave generation is complex, depending on the distance (fetch) over which the wind blows and many other factors, e.g. see [12.21], but will not be discussed here.

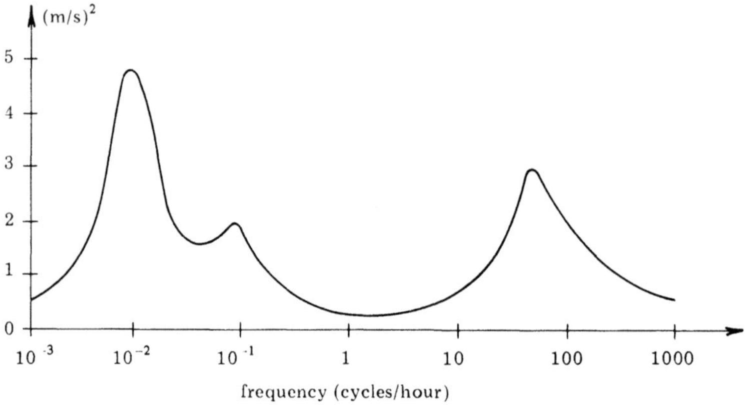

Figure 12.3. Spectrum of horizontal wind speed at Brookhaven [12.37]

Spectral representation: If the variations in the water surface elevation *above mean still water level* are denoted by $\eta(t)$, this quantity may be modelled as a zero-mean, continuous-space, continuous-time stochastic process, which, over short periods of time (say, 6 hours) may be assumed to be stationary. The autocorrelation function of the process is then

$$R_{\eta\eta}(\tau) = E[\eta(t)\eta(t+\tau)] \quad , \quad -\infty \leqslant \tau \leqslant \infty \tag{12.3}$$

and the *mean square spectral density* (or *spectral density*) $S'_{\eta\eta}(\omega)$ is defined as the complex Fourier transform [12.30] of $R_{\eta\eta}(\tau)$ and is given by

$$S'_{\eta\eta}(\omega) = \frac{1}{2\pi}\int_{-\infty}^{\infty} R_{\eta\eta}(\tau)e^{-i\omega\tau}d\tau \quad , \quad -\infty \leqslant \omega \leqslant \infty \tag{12.4}$$

where ω is circular frequency in radians per second.

[Note: For equation 12.4 to apply, the mean value of the process $\mu_\eta(t)$ must be zero].

The associated inverse relationship or *inverse Fourier transform* enables $R_{\eta\eta}(\tau)$ to be recovered from $S'_{\eta\eta}(\omega)$ and is given by

$$R_{\eta\eta}(\tau) = \int_{-\infty}^{\infty} S'_{\eta\eta}(\omega)e^{i\omega\tau}d\omega \quad , \quad -\infty \leqslant \tau \leqslant \infty \tag{12.5}$$

Equations (12.4) and (12.5) are known as a *Fourier transform pair.* $S'_{\eta\eta}(\omega)$, as defined by equation (12.4), is a real, non-negative, even (i.e. $S'_{\eta\eta}(\omega) = S'_{\eta\eta}(-\omega)$) function of ω.

It should be noted that the factor $1/2\pi$ which appears in equation (12.4) is put instead in equation (12.5) by some authors.

A very important property of the spectral density is obtained by combining equations (12.3) and (12.5) and setting $\tau = 0$, namely

$$E[\eta^2(t)] = \int_{-\infty}^{\infty} S'_{\eta\eta}(\omega)d\omega \tag{12.6}$$

i.e. the area under the spectral density curve is equal to the mean square of the process. But, using equation (2.41) it can easily be shown that

$$E[\eta^2(t)] = \mu_\eta^2 + \sigma_\eta^2 \tag{12.7}$$

and, since μ_η is zero, it follows that

$$\int_{-\infty}^{\infty} S'_{\eta\eta}(\omega)d\omega = \sigma_\eta^2 \tag{12.8}$$

where σ_η^2 is the variance of the process.

The form of the spectrum $S'_{\eta\eta}(\omega)$ defined in equation (12.4) is known as a double-sided spectrum and involves the concept of negative as well as positive frequencies. This form arises naturally from the basic definitions, but is not the most helpful for engineering purposes. For the latter, the equivalent one-sided spectrum $S_{\eta\eta}(\omega)$, $0 \leqslant \omega \leqslant \infty$, is generally adopted and will be used here (e.g. figure 12.3).

Clearly,

$$E[\eta^2(t)] = \int_{-\infty}^{\infty} S'_{\eta\eta}(\omega)d\omega = \int_{0}^{\infty} S_{\eta\eta}(\omega)d\omega \tag{12.9}$$

and thus $S_{\eta\eta}(\omega) = 2S'_{\eta\eta}(\omega)$. If the circular frequency ω, measured in radians per second, is replaced by cyclic frequency f, measured in Hertz (cycles per second), then it is easy to show that

$$S_{\eta\eta}(f) = 2\pi S_{\eta\eta}(\omega) = 4\pi S'_{\eta\eta}(\omega) \quad \text{with } \omega = 2\pi f \tag{12.10}$$

[Note: Confusion between single and double-sided spectra and between spectra expressed in Hertz and those in radians/second is a common source of error].

Practical measures of sea-state: The *height* H of a single wave is normally defined as the total range of $\eta(t)$ in the time interval T_0 between two consecutive zero up-crossings by $\eta(t)$, see figure 12.4. T_0 is the corresponding *wave period*, and in general more than one maximum and one minimum of $\eta(t)$ will occur during this time.

The *sea-state* itself is characterized by the distribution of wave heights and periods. Two of the most common practical measures of sea-state are »significant wave height» H_s and »mean zero-crossing period» T_z. H_s is defined as the mean height of the highest one-third of all the waves recorded during a period of observation T, and is sometimes denoted $H_{1/3}$ (or $H^{(1/3)}$). T_z is the mean of the sequence of times T_0 between successive up-crossings of the mean still water level during the same period of observation T. Both H_s and T_z are thus sample statistics relating to the observed sea-state. As might be expected, H_s and T_z are not independent, the larger values of H_s generally being associated with waves of longer period.

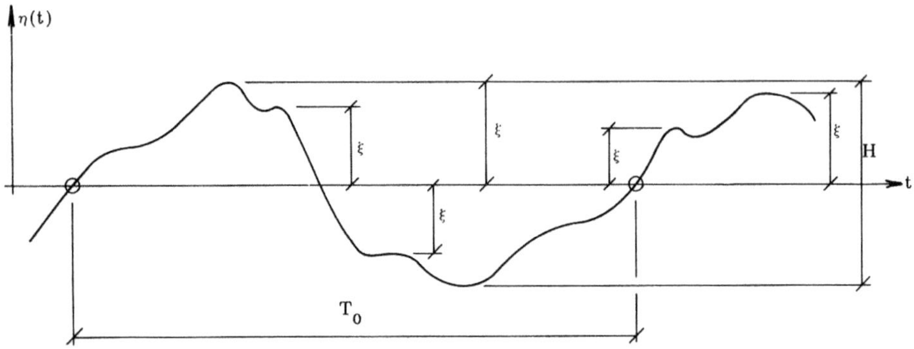

Figure 12.4

Wiegel [12.38] has obtained a regression relationship between log H_s and log T_z based on data from the North Sea and the Gulf of Mexico. When H_s is measured in metres, T_z in seconds was found to be given by

$$T_z = 3.349\, H_s^{0.559} \tag{12.11}$$

However, as shown by Draper and Squire [12.14] in data obtained for a complete year in the North Atlantic, there is considerable scatter in the relationship between H_s and T_z. Houmb and Overvik [12.19] have fitted a two-parameter Weibull distribution to 3925 samples of data obtained from the northern North Sea to determine a conditional distribution of T_z given H_s, as follows

$$F_{T_z | H_s}(t_z | h_s) = 1 - \exp[-(\frac{t_z}{k(h_s)})^{\gamma(h_s)}] \tag{12.12}$$

where

$$k(h_s) = 6.05 \exp(0.07\, h_s)$$

$$\gamma(h_s) = 2.35 \exp(0.21\, h_s)$$

with h_s in metres.

Not surprisingly, there are discrepancies between the results obtained during different investigations, probably due to differing fetches and non-homogeneous wind fields, but for heavy seas (large H_s) the various predictions of T_z are quite close and the conditional variance of T_z reasonably small.

Practical spectral forms for water surface elevation, $\eta(t)$: Research by various oceanographers has led to a number of semi-empirical expressions for the form of the spectra $S_{\eta\eta}(\omega)$ of water surface elevation $\eta(t)$, (generally called wave spectra). Two commonly used spectra are the Pierson-Moskowitz (P-M) [12.31] and the JONSWAP [12.17].

Pierson and Moskowitz found that for fully developed seas (i.e. those that occur when the wind blows uninterrupted over an effectively unlimited fetch) the wave spectra approximate to a single form which is dependent only on wind speed and two dimensionless constants α and β. In its original form, the P-M spectrum was given by

$$S_{\eta\eta}(\omega) = \frac{\alpha g^2}{\omega^5} e^{-\beta(\frac{\omega_0}{\omega})^4} \quad , \quad 0 \leqslant \omega \leqslant \infty \tag{12.13}$$

where

ω is frequency in radians/second

$\omega_0 = g/u_{19.5}$

g is acceleration due to gravity (9.81 m/s²)

$u_{19.5}$ is wind speed in m/s at a height of 19.5 m above mean still water level

α = 0.0081 (Phillip's empirical constant)

β = 0.74

Making use of equation (12.10), the P-M spectrum may be expressed as a function of cyclic frequency f, as follows

$$S_{\eta\eta}(f) = \frac{\alpha g^2}{(2\pi)^4 f^5} \exp(-\beta(\frac{g}{2\pi u_{19.5} f})^4) \quad , \quad 0 \leqslant f \leqslant \infty \tag{12.14}$$

The spectrum is shown in figure 12.5 for various values of wind speed $u_{19.5}$. This illustrates the increase in wave energy with increase in wind speed and the corresponding decrease in the frequency of the spectral peak (increase in dominant wave period). The frequency of the peak f_p is obtained by differentiating $S_{\eta\eta}(f)$ with respect to f and equating to zero, giving

$$f_p = (0.8\,\beta)^{0.25} \frac{g}{2\pi u_{19.5}} \tag{12.15}$$

Finally, combining equations (12.14) and (12.15) gives

$$S_{\eta\eta}(f) = \frac{\alpha g^2}{(2\pi)^4 f^5} \exp(-1.25(\frac{f_p}{f})^4) \quad , \quad 0 \leqslant f \leqslant \infty \tag{12.16}$$

The *JONSWAP* spectrum was developed from the results of the Joint North Sea Wave Project [12.17] and applies to wind generated waves under conditions of limited fetch and homogeneous wind fields. It is given by

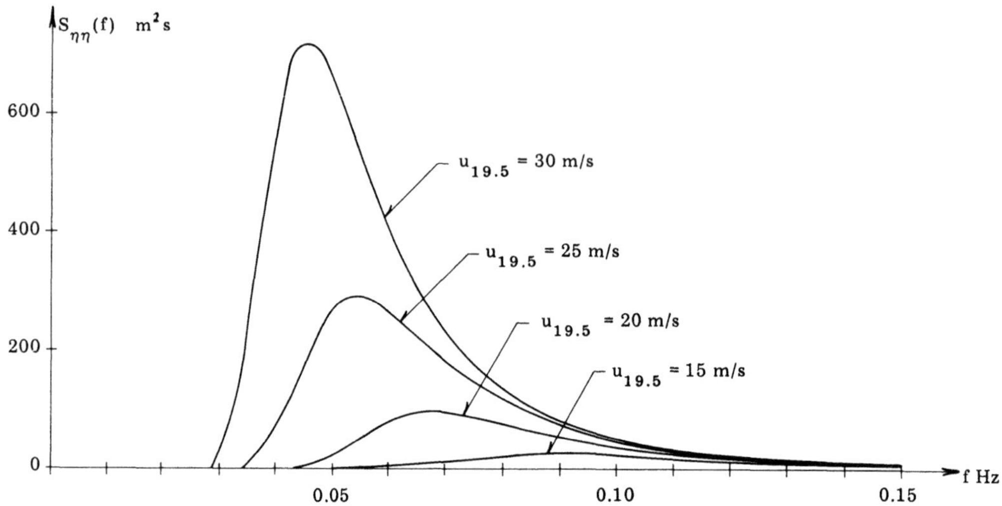

Figure 12.5. Pierson-Moskowitz spectrum for various wind speeds, $u_{19.5}$.

$$S_{\eta\eta}(f) = \frac{\alpha g^2}{(2\pi)^4 f^5} \exp(-1.25 (\frac{f_p}{f})^4) \gamma^a \qquad (12.17)$$

where

$$\alpha = 0.076 \, \tilde{x}^{-0.22} \qquad (12.18)$$

$$f_p = 3.5 \frac{g}{u_{10}} \tilde{x}^{-0.33} \qquad (12.19)$$

$$a = \exp(-\frac{1}{2} (f - f_p)^2 / (\sigma f_p)^2) \qquad (12.20)$$

γ is the peak enhancement factor (typically in the range 3-5)

and where the non-dimensional fetch $\tilde{x} = gx/u_{10}^2$

g is acceleration due to gravity (9.81 m/s²)

x is the true fetch in metres

u_{10} is the wind speed in m/s at a height of 10 m above mean still water level

$\sigma = \sigma_a = 0.07 \qquad f < f_p$

$\sigma = \sigma_b = 0.09 \qquad f > f_p$

The form of equation (12.17) is similar to equation (12.16) but with the addition of a peak enhancement term γ^a. The exponent a, given by equation (12.20), has a maximum value of unity which occurs when $f = f_p$. The peak of the JONSWAP spectrum is therefore γ times the equivalent P-M peak, showing that more energy is concentrated in the narrow band of frequencies around the peak. It will also be seen that the parameters α and f_p are both dependent on the non-dimensional fetch parameter $\tilde{x}$, both decreasing as the non-dimensional fetch is reduced (e.g. if the real fetch x is reduced while the wind speed is kept constant).

For given values of α and f_p, there is greater energy associated with the JONSWAP spectrum than with the P-M spectrum since, with the enhanced peak, the area under the JONSWAP spectrum, and hence $\sigma_\eta^2(t)$ (see equation (12.8)), is larger. However, it should be noted that the two spectra are not directly comparable because one is for fully-developed seas and the other is for considions of limited fetch. A further difference is the height for the reference wind speed. The JONSWAP spectrum, for a fetch of 600 km and various wind speeds, is shown in figure 12.6.

The JONSWAP spectrum has five independent parameters, x, u_{10}, γ, σ_a and σ_b, some of which may be treated as random variables. In the JONSWAP experiment, the shape parameters γ, σ_a and σ_b displayed considerable scatter but, unlike the parameters α and f_p, showed no significant trend with $\tilde{x}$. Their mean values were $\mu_\gamma \simeq 3.3$, $\mu_{\sigma_a} \simeq 0.07$, $\mu_{\sigma_b} \simeq 0.08$. However, the most important quantities are probably u_{10}, x and γ, all of which should be treated as random variables in the reliability analysis of a dynamically sensitive offshore structure. Because fetch is dependent on wind direction for any given offshore location, the information ideally required is the joint probability distribution of extreme wind speed and direction, together with a suitable probability distribution for γ.

Use of parameterized JONSWAP-type spectra: In the preceding paragraphs the Pierson-Moskowitz and JONSWAP spectra have been discussed in the context of finding suitable models for relating extreme wind speed to sea-state, for the purpose of undertaking reliability analyses of dynamically sensitive offshore structures. If the parameters of the probability distributions of

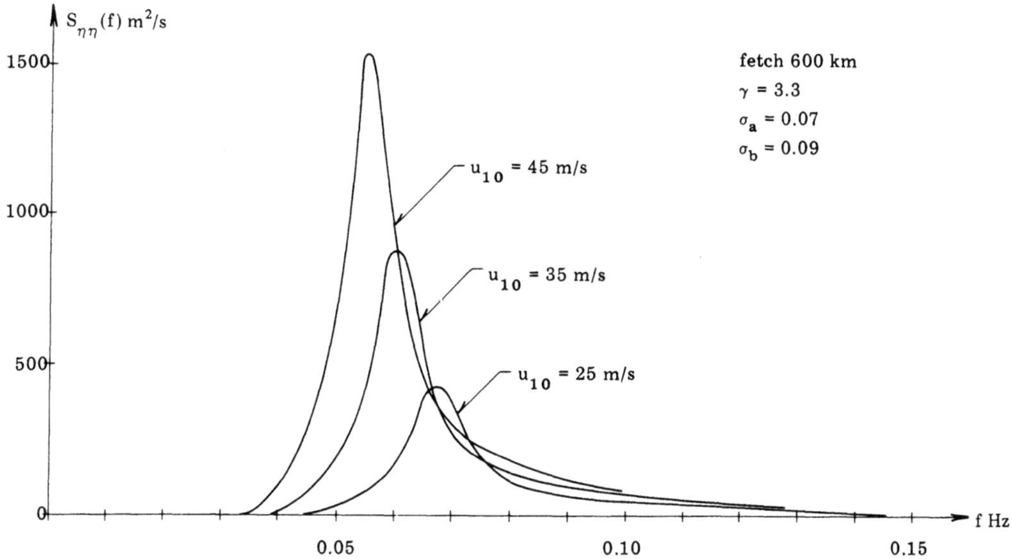

Figure 12.6. JONSWAP spectrum for various wind speeds, u_{10}.

extreme wind speed are known, then this approach can be adopted (see also section 12.3.1). However, in some cases sufficient statistical data on wave heights and periods may be available to allow the reliability analyst to formulate the problem directly in terms of the probability distributions of the sea-state parameters, e.g. H_s and T_z. In such cases, it is necessary to relate the wave spectra to these sea-state parameters.

If $\eta(t)$ is assumed to be a stationary zero-mean Gaussian process (see chapter 9) it has been shown by Rice [12.32] [12.33] that the expected number of up-crossings by $\eta(t)$ through the level zero per unit time is given by

$$\nu_0 = E[N'_+(0)] = (m_2/m_0)^{\frac{1}{2}} \tag{12.21}$$

where m_0 and m_2 are respectively the zero^th and second moments of the spectrum $S_{\eta\eta}(f)$, defined, for the general case, by

$$m_n = \int_0^\infty f^n S_{\eta\eta}(f)df \tag{12.22}$$

Hence, T_z, the mean time between zero up-crossings, is given by

$$T_z = E[T_0] = (E[N'_+(0)])^{-1} = (m_0/m_2)^{\frac{1}{2}} \tag{12.23}$$

The spectral bandwidth parameter ϵ is defined by

$$\epsilon^2 = \frac{m_0 m_4 - m_2^2}{m_0 m_4} \tag{12.24}$$

and lies in the range $0 < \epsilon < 1$. When ϵ is close to zero, the spectrum consists of a narrow band of frequencies and the spectrum is termed *narrow-band*.

Consider now the heights ξ of *all* the wave peaks (maxima in $\eta(t)$) measured from the mean still-water level ($\eta(t) = 0$) - see figure 12.4. Cartwright and Longuet-Higgins have shown (see figure 4 of [12.10]) that for reasonably narrow-band spectra ($\epsilon < 0.4$) the mean height of the highest one-third of all peaks, $\xi_{1/3}$, is given by

$$\xi_{1/3} \simeq 2\sqrt{m_0} \tag{12.25}$$

Furthermore, it can be shown that, for reasonably narrow-band spectra, the largest depth of a trough between zero crossings is almost equal to the height of the wave crest which preceeded it, so that the wave height $H \simeq 2\xi$. Thus the significant wave height H_s may be determined from

$$H_s \simeq 2\xi_{1/3} \simeq 4\sqrt{m_0} \tag{12.26}$$

The last quantity that will be described here is T_p, the expected time between successive peaks (maxima) in the process $\eta(t)$. For a stationary zero-mean Gaussian process, T_p can be shown

[12.9] to be related to m_2 and m_4 by

$$T_p = (m_2/m_4)^{\frac{1}{2}} \tag{12.27}$$

Hence the ratio of the number of peaks to the number zero crossings, which is a measure of the degree of irregularity of $\eta(t)$, is given by

$$r = T_m/T_z = m_2/(m_0 m_4)^{\frac{1}{2}} \tag{12.28}$$

with $0 < r < 1$. For narrow-band spectra r is close to unity, and the number of maxima is not significantly greater than the number of zero up-crossings.

Use may now be made of equations (12.23) and (12.26) to relate the sea-state parameters H_s and T_z to the moments of the spectrum $S_{\eta\eta}(f)$, and hence to obtain expressions for the spectral parameters in terms of H_s and T_z. Houmb and Overvik [12.19] have undertaken such a parameterization of the JONSWAP spectrum (equation (12.17)) to determine the parameters α, γ and f_m for different sets of values of H_s and T_z. This information can be used, together with the appropriate extreme joint distribution function of H_s and T_z for a particular offshore location when this is known, as the input for a structural reliability analysis.

12.2.2 Wave model

Wave theory is a subject in its own right and cannot be discussed in detail in a book on *structural* reliability theory. Attention will be focussed therefore on two-dimensional small-amplitude simple harmonic waves, generally known as Airy waves. They are two-dimensional in the sense that the wave crests are assumed to be parallel and to extend to infinity, and of small amplitude in comparison with wave-length and water depth. Airy waves cannot really be considered to occur in practice (actual deep-water waves correspond more closely to such wave theories as Stokes' 5th order) but are important because of their central role in spectral analysis in which the sea state is represented by an infinite number of Fourier components. For a thorough presentation of Airy and other wave theories the reader should study a specialist text, e.g. Kinsman [12.21].

The assumptions on which linear, or Airy, wave theory are based are as follows:

- the flow is two-dimensional, irrotational and incompressible,
- the pressure at the free surface is negligible,
- the non-linear terms in the equations of motion are negligible compared with the linear terms,
- the water surface displacements are small so that the $(\text{velocity})^2$ terms are negligible.

Given these assumptions, it can be shown [12.21] that the water surface elevation η above still water lavel may be expressed as a function of time t and of horizontal distance x in the direction normal to the crest

$$\eta(x, t) = a\cos(kx - \omega t) \tag{12.29}$$

where

a is the wave amplitude

k is the *wave number* (in rads/m)

ω is the *wave frequency* (in rads/s)

Figure 12.7 shows a cross-section through a typical linear wave. It should be noted that this differs from figure 12.4 in which η is plotted for some *fixed point* as a function of time. Defining the *wave period* $T_0 = 2\pi/\omega$, the *wave length* $L = 2\pi/k$ and the *wave height* $H = 2a$, equation (12.29) may be re-written as

$$\eta(x, t) = \frac{H}{2} \cos 2\pi \left(\frac{x}{L} - \frac{t}{T_0}\right) \tag{12.30}$$

but, because of its greater simplicity, the form of equation (12.29) will be retained for further calculations.

The loading on an offshore jacket structure depends, however, on the water particle velocities $\dot{u}_p$ and accelerations $\ddot{u}_p$. For vertical members it is necessary to know the horizontal components $\dot{u}_p$ and $\ddot{u}_p$ as functions of depth. Taking z as the vertical distance measured upwards from still water level (see figure 12.7), linear wave theory gives the water particle velocities and accelerations as

$$\dot{u}_p(z, x, t) = a\omega \frac{\cosh k(z + d)}{\sinh kd} \cos(kx - \omega t) \tag{12.31}$$

$$\ddot{u}_p(z, x, t) = a\omega^2 \frac{\cosh k(z + d)}{\sinh kd} \sin(kx - \omega t) \tag{12.32}$$

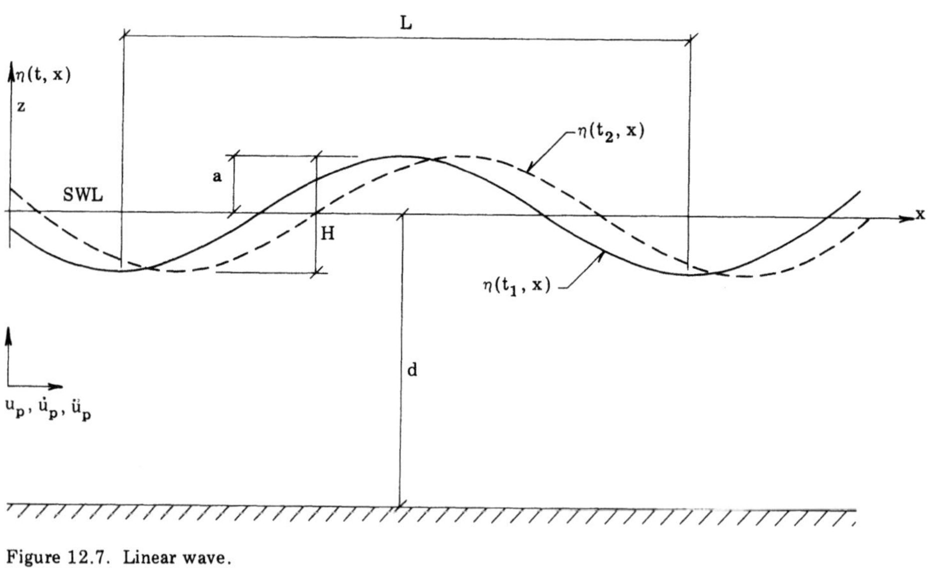

Figure 12.7. Linear wave.

Furthermore, it can be shown that ω, k and d are related by the expression

$$\omega^2 = gk \tanh kd \tag{12.33}$$

which allows the wave number k to be evaluated for known water depth d and any selected frequency component ω. Equations (12.31) and (12.32) are strictly valid only for $-d < z < 0$, but it is possible to apply them for distances z up to the free water surface.

The effect of current: Under extreme storm conditions the predominant loading on offshore structures is that due to waves, but currents are also present and their effect may be significant. In remote offshore locations tidal currents are generally small, but wind generated currents are likely to arise under storm conditions. It is normally assumed that the current velocities at different depths follow a simple power law function of the current velocity $\dot{u}_c(0)$ at the sea surface, giving

$$\dot{u}_c(z) = \dot{u}_c(0)\,((d+z)/d)^{\alpha} \quad , \quad -d \leqslant z \leqslant 0 \tag{12.34}$$

where d is the total water depth and α is a constant.

It is known that wave-current interaction has an effect on hydrodynamic loading, but this is not easy to allow for, and is generally ignored. The structure is therefore analysed simply by using the total particle velocity at each level z, obtained by vectorial addition of the components $\dot{u}_p(z)$ and $\dot{u}_c(z)$.

Ideally, the information required for a structural reliability analysis is the joint distribution function of wind speed, wave direction, current speed and current direction; but this information is likely to be difficult or impossible to obtain for a given offshore location and the analyst may be forced to adopt some simplifying procedures, e.g. analysing the structure first with zero current velocity and then again, assuming that the current and wave directions are the same, to see whether the effect of current is significant.

12.2.3 Loading model

Fluid loading: In section 12.2.1, the concept of a wave spectrum was introduced and ways were described for relating the parameters of the spectrum either to wind speed or to the sea state statistics H_s and T_z. In section 12.2.2, the equations of linear wave theory were given, enabling water particle velocities and accelerations to be determined for components of a wave having amplitude a and frequency ω. In this section, we examine the forces acting on slender members immersed in a fluid which has an instantaneous horizontal velocity $\dot{u}_p$ and acceleration $\ddot{u}_p$.

The most widely accepted approach to the calculation of the wave forces on a vertical slender *rigid* body extending from the sea bed to above the water surface is that due to Morison et al. [12.28]. It is assumed that the wave force can be expressed as the sum of

- a drag force proportional to the square of the water particle velocity, which may be represented by a drag coefficient having substantially the same value as for steady flow, and
- an inertia force proportional to the horizontal component of the accelerative force exerted on the virtual mass of the water displaced by the body.

It should be noted that the virtual mass governing the inertia force is itself composed of two parts, the mass of water displaced by the actual volume of the slender body and an (externally) entrained mass of water which moves with the body. The latter is known as *added mass* and depends on the shape of the body and its surface roughness.

Morison's assumptions can be taken to be valid provided that the body does not significantly affect the wave pattern. This is true if the width of the body D is less than about 20% of the relevant wavelength. The total force at any instant of time t acting on an incremental length of the body dz is then given by

$$dP(t) = (\frac{C_d}{2} \rho D\dot{u}_p|\dot{u}_p| + C_m \rho A\ddot{u}_p)dz \tag{12.35}$$

where

ρ is the water density

$\dot{u}_p$ is the instantaneous water particle velocity normal to the longitudinal axis of the body

$\ddot{u}_p$ is the corresponding water particle acceleration

A is the cross-sectional area of the body

C_d is a drag coefficient

C_m is an inertia coefficient

The total horizontal force on the vertical rigid body can be obtained from

$$P(t) = \int_{z=-d}^{z=0} dP(t) \tag{12.36}$$

To distinguish between the contributions of the actual displaced mass per unit length ρA and the added mass $(C_m - 1)\rho A$, equation (12.35) may be re-written as

$$dP(t) = (\frac{C_d}{2} \rho D\dot{u}_p|\dot{u}_p| + (C_m - 1)\rho A\ddot{u}_p + \rho A\ddot{u}_p)dz \tag{12.37}$$

However, most fixed offshore structures cannot be assumed to be rigid and, indeed, those in deep water may exhibit appreciable dynamic response. In these circumstances, the induced structural velocities $\dot{u}_s$ and accelerations $\ddot{u}_s$ may be significant in comparison with $\dot{u}_p$ and $\ddot{u}_p$ necessitating a modification to equation (12.35). Following the discussion by Chakrabarti [12.11] on a paper by Malhotra and Penzien [12.25] it can be shown that the revised form of Morison's equation to allow for structural motion is

$$dP(t) = (\frac{C_d}{2} \rho D(\dot{u}_p - \dot{u}_s)|\dot{u}_p - \dot{u}_s| + (C_m - 1)\rho A(\ddot{u}_p - \ddot{u}_s) + \rho A\ddot{u}_p)dz \tag{12.38}$$

The main difficulty with the application of Morison's equation, either in its original or revised forms, lies in the proper choice of the coefficients C_d and C_m from a wide range of published data. Although these coefficients can be shown to vary systematically with other parameters such as Reynolds' number, Keulegan-Carpenter number and relative roughness [12.16] there is still considerable residual uncertainty. C_d and C_m may therefore be sensibly treated as random variables and must be included as such in a structural reliability analysis. (See section 12.3.2 for further discussion of C_d and C_m).

Other loading: Apart from the effects of waves and currents discussed above, offshore structures are subject to wind loads, superimposed deck loads, drilling and operational loads, together with both minor and major collision loads. Each needs to be taken into account in a realistic assessment of structural reliability; but relatively simple models can be used for these actions when the overall response is insensitive to them, which is often the case. Major collision loads are of course an exception and require special treatment.

12.2.4 Natural frequency model

The natural frequencies of structures in their various »normal» modes of vibration depend on the spatial distributions of mass and component stiffness. Offshore jacket structures unlike typical building structures contain very few non-structural components, at least below the level of the superstructure. Because of this and because the uncertainty in Young's modulus E for steel is very small, the overall stiffness of such frames can be considered a deterministic function of E and the member dimensions. Furthermore, because the overall stiffness of the structure to horizontal wave loading is a linear function of the stiffness properties of a large number of individual components which for fabrication reasons can be assumed to be independent, the total uncertainty in the stiffness of the jacket itself can be neglected. The major remaining source of uncertainty is in the short-term elastic properties of the soil foundations. A discussion of this is beyond the scope of the present text, but should not be overlooked in practice.

Uncertainties in the mass of an offshore structure arise from the possibility of changes in the superstructure loads during the operating life such as would result from the storage of additional equipment and structural modifications. These uncertainties should also be modelled and their effect on natural frequency taken into account.

12.2.5 Evaluation of structural response

Structural model for loading and spectral analysis: The reliability analysis of an offshore structure requires the repetition of a spectral analysis a number of times, depending on the number of basic variables affecting the response.

For this reason, the mathematical idealisation of the structure that has to be made for the purposes of representing the loading, the damping and the distributed masses should be as simple as possible and yet retain the important structural and hydrodynamic characteristics of the actual jacket.

A jacket structure can be idealised for the purposes of wave loading by a number of vertical members extending from the sea bed to above the water surface and divided into zones each representing members at a particular location in the real structure. The diameter of each section of the idealised model can be chosen such that the frontal area exposed to the waves is equal to the sum of the individual tubular members in that zone. Using this approach, the actual value of the drag coefficient C_d can be used, but it is necessary to adjust the inertia coefficient C_m by the factor $\sum D_i^2/D^2$ where D is the diameter of the equivalent tubular member and D_i are the actual diameters.

For the purposes of determining the dynamic response, a jacket structure may be idealised sufficiently accurately by a number of lumped masses, particularly for the first mode of vibration which dominates the overall response to wave loading.

Spectral analysis: The preceding parts of section 12.2 have provided the basic models for the spectral analysis of a jacket structure under wave and current loading. The input to such calculations is the *spectrum* of water surface elevation $S_{\eta\eta}(\omega)$ and the vertical profile of current velocity. The output from the analysis is a *spectrum* of structural displacements $S_{SS}(\omega)$, or more precisely a spectrum of modal amplitudes in each of the normal modes of vibration which are considered. The complex set of calculations which are required to obtain $S_{SS}(\omega)$ from $S_{\eta\eta}(\omega)$ are summarised in Appendix B.

Making use of the general relationship given in equation (12.8), the variance of the structural displacements σ_S^2 may be obtained from

$$\sigma_S^2 = \int_0^\infty S_{SS}(\omega)d\omega \tag{12.39}$$

where $S_{SS}(\omega)$ is a single-sided spectrum.

It should be emphasised that the final outcome from such a spectral analysis is *probabilistic* in nature in that only the variance (and the mean value) of the displacements are determined. Further analysis is therefore required to obtain suitable peak stresses for use in design or safety checking, as described in section 12.2.6 below. However, it is clear that σ_S^2 *itself* as defined by equation (12.39) is a *deterministic* function of the set of basic random variables $\overline{X}$ (e.g. wind speed) which form the parameters of the input spectrum $S_{\eta\eta}(\omega)$, together with other basic random variables such as drag and inertia coefficients. We may therefore write

$$\sigma_{S|\bar{x}}^2 = f(S_{\eta\eta}(\omega)|\bar{x}) \tag{12.40}$$

This provides the basis for creating a failure function of the form described in chapters 4, 5 and 6 suitable for use in a level 2 reliability analysis.

12.2.6 Evaluation of peak response

It was assumed in sections 12.2.1 and 12.2.2 that the water surface elevation $\eta(t)$ can be sensibly modelled for short periods of time as a stationary zero-mean Gaussian process. However, because of the non-linear drag term ($C_d\,\rho D\dot{u}_p|\dot{u}_p|/2$) in Morison's equation, the final structural response will be non-Gaussian. As discussed in Appendix B, an approximation to the true response variance can be obtained by a method of equivalent linearisation in which the root mean square error of the linearised response is minimised.

The response may therefore be interpreted as an equivalent Gaussian process, with $S_{SS}(\omega)$ being the spectral density of the resulting displacements S(t) (e.g. the amplitude of the first mode of vibration).

From a knowledge of $S_{SS}(\omega)$ it is possible to determine the peak response of the structure during an exposure time T. Making use of equation (12.27), the expected number of maxima in S(t) occurring during time T is

$$N = \frac{T}{T_p} = T(m_4/m_2)^{\frac{1}{2}} \tag{12.41}$$

where (see equation (12.22)) the spectral moments m_2 and m_4 are defined by

$$m_n = \int_0^\infty f^n S_{SS}(f)df = \frac{1}{(2\pi)^n} \int_0^\infty \omega^n S_{SS}(\omega)d\omega \tag{12.42}$$

It is then necessary to find the distribution function of the maximum S_e of the N individual maxima of S(t) that occur during time T. This has been shown by Davenport [12.12] to be

$$F_Z(z) = \exp(-\nu_0 T \exp(-\frac{z^2}{2})) \tag{12.43}$$

where

$$Z = S_e/\sigma_S \tag{12.44}$$

and $\nu_0 = (m_2/m_0)^{\frac{1}{2}}$ as defined in equation (12.21).

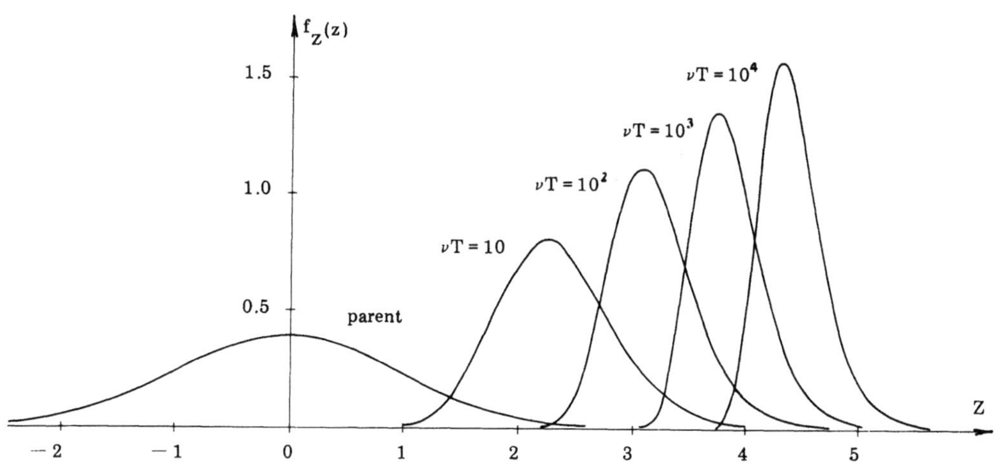

Figure 12.8.

The density function f_Z is plotted in figure 12.8 for various large values of νT. The mean and standard deviation of Z are given by

$$\mu_Z = (2 \ln \nu_0 T)^{\frac{1}{2}} + \frac{\gamma}{(2 \ln \nu_0 T)^{\frac{1}{2}}} - \frac{\pi^2/6 + \gamma^2}{2(2 \ln \nu_0 T)^{\frac{3}{2}}} \tag{12.45}$$

$$\sigma_Z = \frac{\pi}{\sqrt{6}} \frac{1}{(2 \ln \nu_0 T)^{\frac{1}{2}}} \tag{12.46}$$

where $\gamma = 0.5772$ (Euler's constant).

Finally, the random variable S_e representing the peak modal displacement is obtained from

$$S_e = Z\sigma_S \tag{12.47}$$

In deterministic design procedures it is common to choose the »peak factor» $Z = \mu_Z$ so that

$$S_e = \mu_Z \sigma_S \tag{12.48}$$

However, in a structural reliability analysis it is possible to treat S_e as an additional basic variable which contributes to the overall uncertainty.

12.2.7 Other models

In sections 12.2.1 to 12.2.6 various mathematical models have been given which provide the structural and reliability analyst with a method of computing the probability distribution of the peak modal displacement of a steel jacket structure in a given period of exposure to a stationary (strictly ergodic) random sea.

As discussed in section 12.2.4, the uncertainties in the stiffness properties of jacket structures can be assumed to be small in comparison with strength properties and up to the levels of load at which the first structural component fails it is reasonable to assume that the overall behaviour of the structure remains linear. For any particular value of peak displacement, the forces and moments may be computed for any part of the structure by conventional methods of linear analysis. This information can then be used in conjunction with appropriate deterministic models for the strength of individual structural components to assess whether or not they will fail.

Finally, with knowledge of the probability distributions of the relevant basic variables and model uncertainties, it is possible to determine the failure probability for each structural component for any given sea state and, by integrating over all possible sea states, to evaluate the total failure probability for any component for the design life of the structure or for any other period of time. Some results for calculations of this type are given in section 12.5.

12.3 PROBABILITY DISTRIBUTIONS FOR IMPORTANT LOADING VARIABLES

Attention will be restricted here to the modelling of three quantities which are of particular importance in the reliability analysis of jacket structures - wind speed and Morison's coefficients C_d and C_m.

12.3.1 Wind speed

For the deterministic design and analysis of structures it is common practice to use a quantity such as the 50-year return wind speed (the wind speed that has an assumed probability of being exceeded of 0.02) as a fixed deterministic value; and considerable effort has been expended in determining this statistic for both on-shore and off-shore areas. However, this is not of direct relevance in a reliability assessment.

In section 12.2.1 it was seen that the input to a stochastic dynamic reliability analysis of a jacket structure could be either the probability distribution of extreme wind speed *or* the joint extreme distribution of the parameters H_s and T_z - in both cases the distributions corresponding to the extreme conditions in the reference period for which the reliability is to be determined (e.g. 25 years). Which approach is used in practice must be governed by the availability of data.

Reliable wind records have been obtained at exposed coastal stations for a number of decades and there are often advantages to be gained in predicting the distributions of extreme sea-states from these wind records, rather than trying to use short-term wave statistics obtained from a particular offshore location. However, as more wave data are obtained the balance may change. Only wind speed is considered here.

As discussed by Davenport [12.13], a simple model of the atmospheric boundary layer which assumes that large scale turbulence is composed of uncorrelated Gaussian fluctuations of wind speed from orthogonal directions leads to a Rayleigh distribution (see page 27) for the mean-hourly wind speed at a particular location. The Rayleigh distribution is a special case of the 2-parameter Weibull distribution and it is found that both long and short-term continuous records of mean-hourly and mean-daily wind speed are accurately modelled by the latter - see figure 12.9.

As stated in chapter 3, the 2-parameter Weibull distribution has a distribution function (see page 43)

$$F_X(x) = 1 - e^{-(\frac{y}{k})^\beta} \quad , \quad y > 0 \tag{12.49}$$

If it is assumed that the parent mean-hourly wind speed X follows this distribution, the maximum of n independent samples taken from the parent will approach a type I asymptotic distribution of the largest extreme, since the Weibull distribution has an upper tail which falls off in a negative exponential manner (see page 40).

As there are 8760 hours in a year, it is a reasonable assumption that the annual maximum mean-hourly wind speed Y is type I maxima (Gumbel) distributed (see page 40), i.e.

$$F_Y(y) = \exp(-\exp(-\alpha(y-u))) \quad , \quad -\infty \leqslant y \leqslant \infty \ , \ \alpha > 0 \tag{12.50}$$

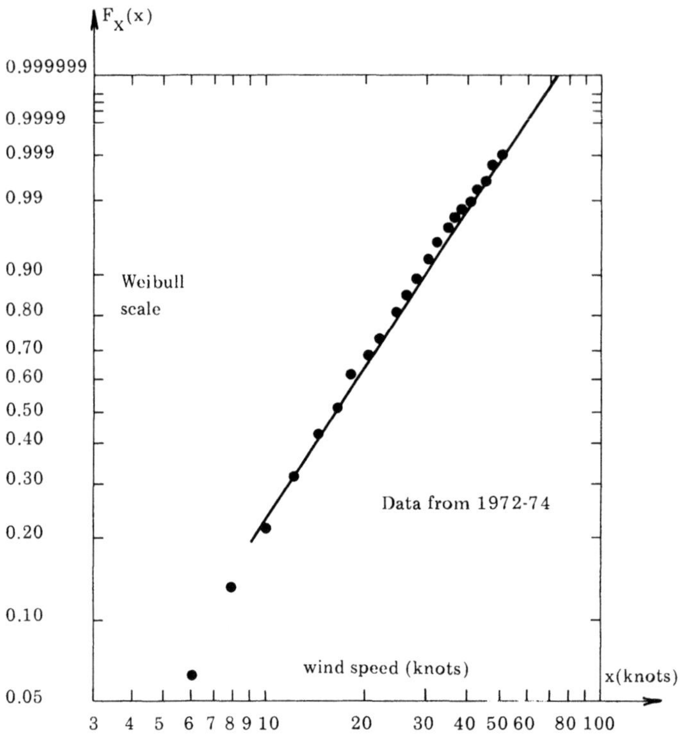

Figure 12.9. Mean-daily wind speed from M.V. Famita (57° 30′N, 3° E)

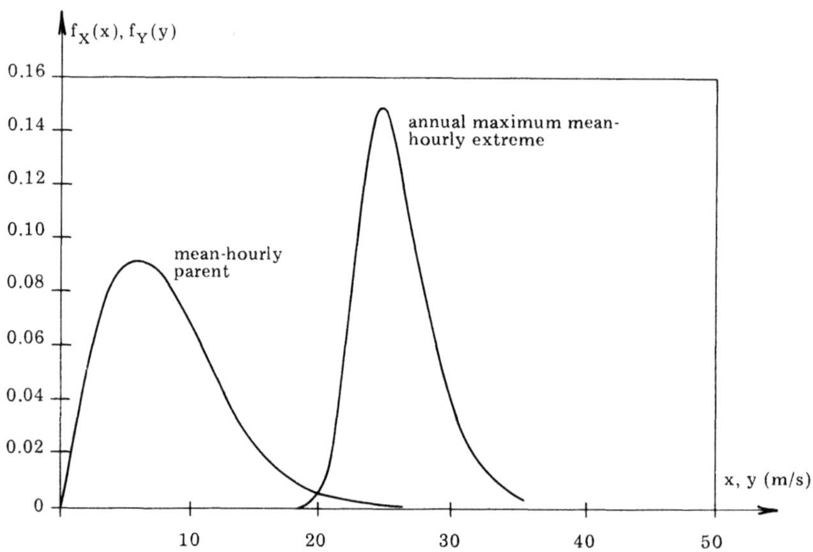

Figure 12.10. Distributions of parent and annual extreme winds for Lerwick.

Taking the useful life of an offshore platform as 25 years, the assumption that the 25 year maximum mean-hourly wind speed has this distribution is even more reasonable, unless there are other influences which begin to dominate at these low probability levels. Such an influence could be the presence of a number of different storm generating mechanisms (e.g. tropical cyclones in addition to fully developed pressure systems). However, for areas such as the North Sea, the major storm generating mechanisms can be assumed to be of a single type.

Analysis of wind data from a large number of meteorological stations, e.g. [12.22], shows that although the mean annual maximum mean-hourly wind speed may vary from one location to another, the coefficient of variation of this quantity is sensibly constant for locations of similar exposure. Taking Lerwick to be reasonably representative of the northern North Sea, a typical value for the coefficient of variation of annual maximum mean-hourly extremes is 12-13%. The distribution of annual maximum values for Lerwick is shown in figure 12.10.

If annual maxima values of mean-hourly wind speed are assumed to be type I maxima distributed, then the probability distribution of extreme values corresponding to larger periods of time may be determined from equation (3.15).

In the preceding paragraphs reference is made to mean-hourly and mean-daily wind statistics. However, whereas most wind data are recorded in terms of 3-second gusts or as mean-hourly (time-averaged) values, most major storms last from 6-12 hours. Conversion factors are available to convert from mean-hourly values to speeds averaged over a number of hours - e.g. [12.9] - but some uncertainty must be attached to these values because the factors that are appropriate for moderate extremes may not be exact for severe extremes.

12.3.2 Morison's coefficients

The Morison force coefficients C_d and C_m (see equations (12.35) and (12.36)) have been shown by experiment to exhibit considerable scatter. A critical evaluation of all the available information has been made by the British Ship Research Association [12.8].

Under uniform conditions with steady flow, constant temperature, viscosity, roughness and geometry the drag coefficient C_d is (probably) a deterministic quantity and considerable experimental work has been undertaken to determine values; but because of the difficulties of scaling, experimental results obtained at low Reynolds' numbers on small models cannot easily be applied to full scale structures with flow at high Reynolds' numbers.*)

Under typical offshore conditions, however, there is oscillating flow with orbital water-particle motion. Work by Sarpkaya [12.35], using a large U-shaped vertical water tunnel, allowing the attainment of flow at high Reynolds' numbers has shown that, for smooth cylinders, the drag and inertia coefficients depend on both the Reynolds' number Re and the Keulegan-Carpenter number N_{KC}.*) For rough cylinders, C_d and C_m have values which become nearly independent of Reynolds' number for values of Re in excess of about 4×10^5, but which depend on Keulegan-Carpenter number and relative roughness k/D. Although the values of C_d obtained by Sarpkaya for smooth cylinders in oscillating flow are about the same as the values obtained for smooth cylinders in steady flow (namely, between 0.6 and 0.7), the values obtained for rough cylinders in oscillating flow range from 1.2 to 1.8 depending on relative roughness. These values are very high and cast some doubt on the values of about 0.7 which are normally used in design. However, the conditions under which these experiments were conducted are very dissimilar to the wave conditions experienced in the open sea.

*) $Re = \frac{\dot{u}_m D \rho}{\nu}$ and $N_{KC} = \frac{\dot{u}_m T}{D}$ where $\dot{u}_m$ is the amplitude of horizontal oscillatory velocity, D is the cylinder diameter, T is the wave period, ρ is the density of water and ν is its viscosity.

In contrast to the above are some results obtained by Exxon on a full-scale section of a 325 mm diameter pile exposed to wave loading, as reported by Kim and Hibbard [12.20]. This permitted an assessment of Morison's coefficients, independent of any wave theory. The mean value obtained for C_d was 0.61 with a coefficient of variation of 0.24, whereas for C_m, the corresponding mean and coefficient of variation were 1.20 and 0.22. No significant differences were found for seas with significant wave heights ranging from 0.8 to 3.0 m.

Experimental data give no clear indication of what are appropriate values to take for C_d and C_m for offshore applications. Although these force coefficients may well be complex deterministic functions of basic parameters such as viscosity, density and flow geometry, these relationships are not known. What is more, even if they were known for simple isolated piles, interaction effects between individual piles in a jacket, the effects of varied marine growth and the interaction of waves and currents would render any simple theory invalid when applied to real structures. There is therefore very considerably uncertainty. In the circumstances it is appropriate to consider C_d and C_m as random variables where the total uncertainty arises from the combined effect of a large number of mainly unknown influences. Mean values for C_d and C_m of 0.75 and 1.8, respectively, with coefficients of variation of between 25% and 35% provides a reasonable probabilistic model for these force coefficients for members of typical jacket structures. The coefficients can be sensibly assumed to be normally distributed.

12.4 METHODS OF RELIABILITY ANALYSIS

12.4.1 General

The basic principles and methods of structural reliability theory have been discussed in chapters 4, 5 and 6, but a number of additional topics need to be considered here in connection with offshore structures.

In chapters 4 and 5 it was seen that the key step in any analysis is the identification of a suitable mathematical model which defines survival in terms of a function

$$M = f(X_1, X_2, \ldots, X_n) = f(\overline{X}) \leqslant 0 \tag{12.51}$$

where $\overline{X}$ is a vector of basic random variables (e.g. wind speed, material properties, dimensions, etc.). In most of the simple structural examples given in earlier chapters, the form of the failure function f is explicit and is easily determined, as for example in the case of the propped cantilever beam given in example 5.3. However, for complex structural systems this is not the case. For dynamically sensitive offshore structures the function f incorporates all the models for loading and response given in section 12.2, together with the failure criterion for the structural component under consideration, e.g. the buckling of a stiffened tubular member under the combined action of end loads and moments. The only sensible form of expression for f in this case is a computer program.

Regardless of the complexity of the function f, the principles of reliability analysis are the same, provided that the basic random variables do not have parameters which themselves change with time. As shown in equation (4.42) the reliability is given by

$$\mathcal{R} = 1 - P_f = 1 - \iint\limits_{f(\bar{X}) \leq 0} \cdots \int f_{X_1, X_2, \ldots, X_n}(x_1, x_2, \ldots, x_n)\, dx_1 dx_2 \ldots dx_n \tag{12.52}$$

where $f_{X_1, X_2, \ldots, X_n}(x_1, x_2, \ldots, x_n)$ is the joint probability density function for the n variables X_i.

As discussed in chapter 4, various methods can be used to determine $\mathcal{R}$ from equation (12.52) when the failure function f is of a simple form and when n is small. However, for complex failure functions involving spectral analysis only level 2 methods are seriously worth considering. Exactly the same theory can be followed as in chapters 4 and 5, but an algorithm more suitable for computer-based calculations is given below. This can be used when the failure criterion cannot be expressed explicitly in terms of the basic variables.

12.4.2 Level 2 method

The level 2 method in its most general form may be interpreted as comprising the following steps:

1) specification of a failure function f in terms of n basic random variables $\bar{X}$
2) creation of a hyper-surface (known as the failure surface) in the n-dimensional space of the basic variables $\bar{X}$ (x-space) by setting $f(\bar{X}) = 0$
3) defining the joint density function $f_{\bar{X}}$ for the n basic variables
4) mapping the failure surface in $\bar{X}$-space to the space of n independent standard normal variables $\bar{Z}$ (z-space)
5) finding the shortest distance, β, from the origin to the failure surface.

For linear failure functions and for sets of basic variables which are jointly-normal, it has been stated in chapter 5 that the following exact relationship holds

$$\mathcal{R} = 1 - P_f = 1 - \Phi(-\beta) \tag{12.53}$$

In nearly all practical problems, however, these conditions are not satisfied and the relationship is only approximate. Nevertheless, the errors are generally very small - see, for example, [12.3].

In chapter 6, it was shown how correlation between basic variables can be taken into account and how the level 2 method can be modified to allow for uncorrelated non-normal basic variables. In the following, it is suggested how the analysis can be undertaken if information is available on the complete joint density function $f_{\bar{X}}$.

General method: As pointed out by Hohenbichler and Rackwitz [12.18], it has been shown by Rosenblatt [12.34] that if $\bar{X}$ is a random vector with a continuous distribution function $F_{\bar{X}}$ then if

$$\begin{aligned} u_1 &= F_1(x_1) \\ u_2 &= F_2(x_2|x_1) \\ &\vdots \\ u_n &= F_n(x_n|x_1, x_2, \ldots, x_{n-1}) \end{aligned} \qquad (12.54)$$

the variables $U_1, U_2, \ldots, U_n$ are uniformly and independently distributed. The functions F_i, $i = 1, 2, \ldots, n$ are defined in (12.56). Hence, if z_i and u_i are related by

$$z_i = \Phi^{-1}(u_i) = \Phi^{-1}(F_i(x_i|x_1, x_2, \ldots, x_{i-1})) \quad , \quad i = 1, 2, \ldots, n \qquad (12.55)$$

the variables $Z_1, Z_2, \ldots, Z_n$ are independent standard normal.

The conditional distribution functions $F_i(x_i|x_1, x_2, \ldots, x_{i-1})$ in equation (12.54) are given by

$$F_i(x_i|x_1, x_2, \ldots, x_{i-1}) = \frac{\int_{-\infty}^{x_i} f_i(x_1, x_2, \ldots, x_{i-1}, s_i)\,ds_i}{\int_{-\infty}^{\infty} f_i(x_1, x_2, \ldots, x_{i-1}, s_i)\,ds_i} \qquad (12.56)$$

where the partial joint density functions f_i, $i = 1, 2, \ldots, n$ are defined by

$$f_i(x_1, x_2, \ldots, x_i) = \int_{-\infty}^{\infty} \cdots \int_{-\infty}^{\infty} f_{\overline{X}}(x_1, x_2, \ldots, x_i, s_{i+1}, \ldots, s_n)\,ds_{i+1} \ldots ds_n \qquad (12.57)$$

and where $f_{\overline{X}}$ is the joint density function of the original set of basic variables.

The inverse relationships corresponding to equations (12.54) and (12.55) are given by

$$\begin{aligned} x_1 &= F_1^{-1}(\Phi(z_1)) \\ x_2 &= F_2^{-1}(\Phi(z_2)|x_1) \\ &\vdots \\ x_n &= F_n^{-1}(\Phi(z_n)|x_1, x_2, \ldots, x_{n-1}) \end{aligned} \qquad (12.58)$$

Finally, substituting equations (12.58) in equation (12.51) the safety margin M may be expressed as

$$M = f(F_1^{-1}(\Phi(Z_1)), F_2^{-1}(\Phi(Z_2)|F_1^{-1}(\Phi(Z_1)), \ldots) \qquad (12.59)$$

Equation (12.59) gives M as a (generally non-linear) function of n independent standard normal variables for which the method of analysis given in section 4 of chapter 5 is directly applicable. The reliability index β is given by

$$\beta = \min_{\bar{z} \in \partial\omega} \left(\sum_{i=1}^{n} z_i^2 \right)^{\frac{1}{2}} \tag{12.60}$$

where $\partial\omega$ is the failure surface in $\bar{Z}$-space.

The fundamental problem with the preceding formulation is that the joint density function $f_{\bar{X}}$ is rarely known and therefore this approach cannot be used. However, in many cases the components of the random vector $\bar{X}$ may be assumed to be stochastically independent in which case equations (12.58) become

$$x_i = F_i^{-1}(\Phi(z_i) \quad , \quad i = 1, 2, \ldots, n \tag{12.61}$$

and

$$M = f(F_1^{-1}(\Phi(Z_1)), \ldots, F_n^{-1}(\Phi(Z_n))) = g(Z_1, Z_2, \ldots, Z_n) = g(\bar{Z}) \tag{12.62}$$

Computational procedure: It is assumed that a failure function g of the form given by equation (12.62) has been obtained for a particular structural problem and that $\bar{Z}$ is a vector of n independent standard normal variables.

Let $\bar{z}^* = (z_1^*, z_2^*, \ldots, z_n^*)$ be the coordinates of the closest point on the failure surface to the origin. Then expanding g as a Taylor series about $\bar{z}^*$

$$g(\bar{z}^* + \delta\bar{z}) \doteq g(\bar{z}^*) + \sum_{i=1}^{n} a_i \delta z_i \tag{12.63}$$

where

$$a_i = \left. \frac{\partial g}{\partial z_i} \right|_{\bar{z} = \bar{z}^*} \tag{12.64}$$

But for small $\delta\bar{z}$

$$g(\bar{z}^* + \delta\bar{z}) \doteq g(\bar{z}^*) = 0 \tag{12.65}$$

and thus

$$\sum_{i=1}^{n} a_i \delta z_i \doteq 0 \tag{12.66}$$

However, θ, the distance from the origin to a random point $\bar{z}$ on the failure surface is given by

$$\theta = \left(\sum_{i=1}^{n} z_i^2 \right)^{\frac{1}{2}} \quad , \quad \bar{z} \in \partial\omega \tag{12.67}$$

and for θ to be a minimum (or maximum)

$$\sum_{i=1}^{n} \frac{\partial \theta}{\partial z_i} \delta z_i = \sum_{i=1}^{n} \frac{z_i}{\left(\sum_{i=1}^{n} z_i^2\right)^{\frac{1}{2}}} \delta z_i = 0 \tag{12.68}$$

If θ *is* a minimum, and assuming that only one minimum exists,

$$\theta = \beta = \min_{\bar{z} \in \partial\omega} \left(\sum_{i=1}^{n} z_i^2\right)^{\frac{1}{2}} = \left(\sum_{i=1}^{n} z_i^{\star 2}\right)^{\frac{1}{2}} \tag{12.69}$$

giving

$$\sum_{i=1}^{n} \frac{z_i^\star}{\beta} \delta z_i = 0 \tag{12.70}$$

Comparison of equations (12.66) and (12.70) shows that for any arbitrary set of small values $\delta \bar{z}$, a necessary condition for them both to be satisfied and for $\bar{z}^\star$ to be the closest point is

$$a_i = c\, z_i^\star / \beta \quad , \quad i = 1, 2, \ldots, n \tag{12.71}$$

where c is a non-zero constant.

For the purposes of calculating β, let $\bar{z}^0$ be an estimate of the coordinates of the closest point and let $(\bar{z}^0 + \delta \bar{z})$ be a better estimate. Then, by expanding g as a Taylor series about the point $\bar{z}^0$ we obtain, as before

$$g(\bar{z}^0 + \delta \bar{z}) \simeq g(\bar{z}^0) + \sum_{i=1}^{n} a_i^0 \delta z_i \tag{12.72}$$

where

$$a_i^0 = \left.\frac{\partial g}{\partial z_i}\right|_{\bar{z} = \bar{z}^0} \tag{12.73}$$

Assuming now that $(\bar{z}^0 + \delta \bar{z}) \simeq \bar{z}^\star$ and that $a_i^0 \simeq a_i$, $i = 1, 2, \ldots, n$, it can be seen that the left hand side of equation (12.72) is approximately zero, giving

$$g(\bar{z}^0) + \sum_{i=1}^{n} a_i^0 \delta z_i \simeq 0 \tag{12.74}$$

and from equation (12.71)

$$z_i^\star \simeq z_i^0 + \delta z_i \simeq a_i^0 \beta / c \quad , \quad i = 1, 2, \ldots, n \tag{12.75}$$

Substituting equation (12.75) in (12.74)

$$g(\bar{z}^0) + \sum_{i=1}^{n} a_i^0 (a_i^0 \beta / c - z_i^0) \simeq 0 \tag{12.76}$$

and thus

$$\beta/c = \frac{\sum_{i=1}^{n} a_i^0 z_i^0 - g(\bar{z}^0)}{\sum_{i=1}^{n} (a_i^0)^2} \tag{12.77}$$

Finally, re-substituting β/c in equation (12.75) gives

$$z_i^\star \simeq a_i^0 \frac{\sum_{i=1}^{n} a_i^0 z_i^0 - g(\bar{z}^0)}{\sum_{i=1}^{n} (a_i^0)^2} \quad , \quad i = 1, 2, \ldots, n \tag{12.78}$$

and from equation (12.69)

$$\beta \simeq \left(\sum_{i=1}^{n} (z_i^\star)^2\right)^{\frac{1}{2}} \tag{12.79}$$

Equations (12.73) and (12.78) together with equation (12.79) provide the basis for an efficient iterative method for calculating β. The procedure is as follows:

1) Select a trial vector $\bar{z}^0$, preferably in the region of $\bar{z}^\star$.

2) Determine a_i^0, i = 1, 2,....,n from equation (12.73).

3) Determine a better estimate of $\bar{z}^\star$ from equation (12.78).

4) Repeat 2) and 3) to achieve convergence.

5) Evaluate β from equation (12.79).

In most practical situations the form of the failure function g will be such that it is not explicitly differentiable. In this case, it is necessary to use numerical differentiation, but this rarely causes difficulties. It should be noted that the sensitivity factors α_i (discussed in chapters 5 and 6) are not directly calculated using the above procedure, but may be evaluated from

$$\alpha_i = \bar{z}^\star/\beta \quad , \quad i = 1, 2, \ldots, n \tag{12.80}$$

Finally, a caution about local minima. Depending on the form of the failure function g, there may or may not be local minima (and local maxima) present. Convergence to one of these points would result in a dangerous over-estimation of the reliability index and hence $\mathcal{R}$. In case of doubt, the analysis can be started using a range of different trial vectors $\bar{z}^0$ and searches for the failure surface can be made in a number of predetermined directions. These problems tend not to arise in simple structural problems, but for complex structures more care has to be taken.

12.5 SOME RESULTS FROM THE STUDY OF A JACKET STRUCTURE

In the last part of this chapter which is based on [12.2], some results are given from the analysis of a typical deep-water jacket structure. The purpose of this is to illustrate the use of the various models and calculation procedures that have been described in earlier sections. It is not intended as a critical review of the reliability of jacket structures which is a subject beyond the scope of the present text.

Figure 12.11 shows the elevations of a structure intended for barge launching at a location in the North Sea with a mean water depth of 156 m. It was designed in accordance with the rules of the American Petroleum Institute [12.1] with full allowance for dynamic response, and was used as a basis for a sensitivity study [12.23], [12.2]. The structure was subjected to a full structural analysis to obtain natural frequencies, mode shapes and influence coefficients for forces and moments in the various members. This detailed deterministic analysis was necessary to provide input for the simplified models used in the reliability analysis and to interpret the response in terms of stresses, as discussed in section 12.2.

The reliability analysis was undertaken to check the safety of the structure under the action of wind, wave and current loading and to assess the sensitivity of the design to the various random variables that affect its behaviour. Use was made of the JONSWAP spectrum to relate extreme wind speed to sea-state.

For present purposes, only results for a single tubular member are discussed (indicated by the arrow in figure 12.11). This leg member acts as a strut carrying a combination of axial load and moment, and failure was deemed to occur at the collapse load predicted by the API design rules [12.1], but treating the dimensions and material properties as random variables and setting the permissible stress equal to the yield stress.

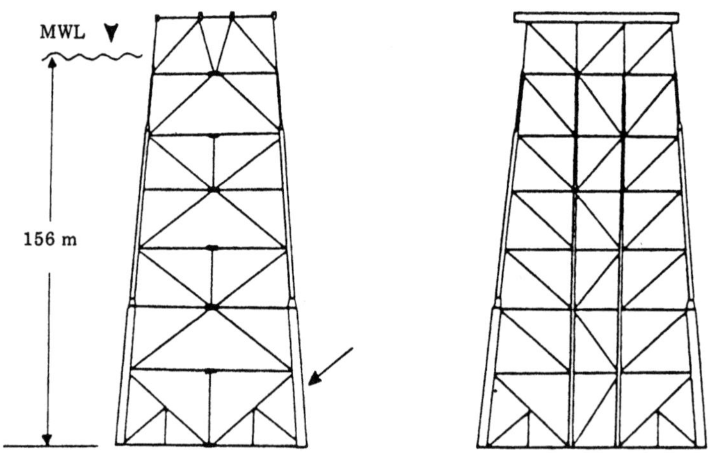

Figure 12.11.

Variable	Distribution	μ_0	σ_0	α
Annual extreme 6-hourly wind	Type I extreme	26.1 m/s	3.3	+ 0.900
Yield stress	Log-normal	380 N/mm^2	18.0	− 0.214
Current speed	Normal	0 m/s	0.14	+ 0.203
C_d	Normal	0.75	0.225	+ 0.199
Marine growth at m.w.l.	Normal	0.16 m	0.06	+ 0.154
Strength model uncertainty	Normal	0.067	0.023	− 0.099
C_m	Normal	1.80	0.27	+ 0.063
Deck load	Normal	24000 t	720	+ 0.047
Thickness of leg wall	Normal	33.5 mm	0.34	− 0.046
Leg diameter	Normal	4191 mm	16.8	− 0.018
Damping ratio	Normal	0.03	0.01	− 0.001
Others	-	-	-	(0.146)

Table 12.1. Analysis of jacket structure.

The parameters of the probability distributions of the important basic variables are given in table 12.1. Using this data, it was found that failure of the member occurs with a probability of about 3.0×10^{-7} in a reference period of 25 years, but no particular significance should be attached to this number. What are of interest, however, are the parameters α_i, which indicate the relative sensitivity of the failure probability to the various basic variables. These are ranked in order of decreasing importance in table 12.1. The fact that wind speed is dominant is not surprising because it is the main loading variable and because its extreme value is subject to considerable uncertainty. The relative importance of C_d compared with C_m is also not unexpected . The contribution of the geometrical variables to the total uncertainty is negligible.

An alternative approach in determining sensitivity is to examine the effects of moderate changes in the distribution parameters. This is illustrated for changes in standard deviation σ_X in figure 12.12a, and for changes in mean μ_X in figure 12.12b. These show that the failure probability is extremely sensitive to the variance of the extreme wind speed and to the mean value of the yield stress. The effect of marine growth is relatively unimportant.

Some further results using a more sophisticated shell analysis for the stiffened cylindrical leg members, in place of the API failure criterion, have been reported in [12.2]. This has enabled important conclusions to be drawn about the relative importance of geometrical imperfections in governing failure.

An immediate field of application for the type of analysis described above is to assist with decision making in areas such as the selection of optimal fabrication tolerances and in the scheduling of cleaning programmes for the removal of marine growth. The main future use, how-

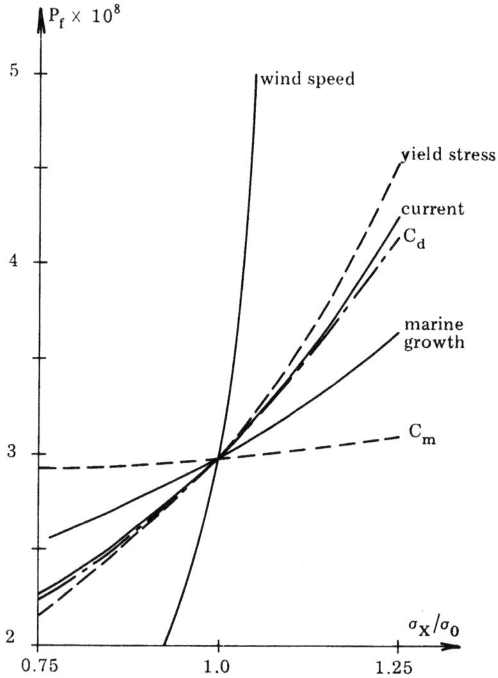

Figure 12.12a.

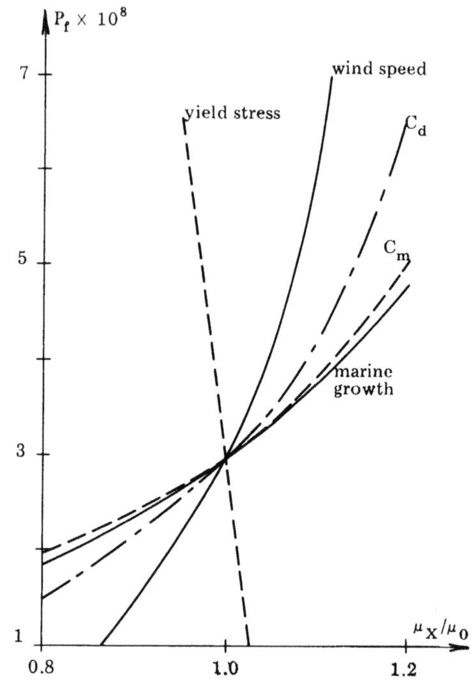

Figure 12.12b.

ever, will be in the reliability assessment of complete structural systems. The development of the close bounds discussed in chapter 8 has brought the goal of complete system reliability analysis in sight, but it would be irresponsible to pretend that, with the present state of knowledge, the problems of undertaking a complete analysis of a complex structure involving many failure modes are not formidable.

BIBLIOGRAPHY

[12.1] American Petroleum Institute: *Recommended Practice for Planning, Designing and Constructing Fixed Offshore Platforms.* API RP 2A Seventh Edition, Jan. 1976.

[12.2] Baker, M. J. and Ramachandran, K.: *Reliability Analysis as a Tool in the Design of Fixed Offshore Platforms.* The Integrity of Offshore Structures. Proc. 2nd International Symposium, Glasgow, July 1981. Applied Science Publishers, 1981.

[12.3] Baker, M. J. and Wyatt, T.: *Methods of Reliability Analysis for Jacket Platforms.* Proc. 2nd International Conference on the Behaviour of Offshore Structures, London, Aug. 1979.

[12.4] Battjes, J. A.: *Probabilistic Aspect of Ocean Waves.* Safety of Structures under Dynamic Loads (ed. Holand, I. et al.) Volume I, Tapir, Trondheim, 1977.

[12.5] Borgman, L. E.: *The Spectral Density for Ocean Wave Forces.* Coastal Eng. Conf., ASCE, 1965.

[12.6] Borgman, L. E.: *A Statistical Theory for Hydrodynamic Forces on Objects.* Hydraulic Engineering Laboratory Wave Research Projects. University of California, Berkeley, 1965.

[12.7] Brebbia, C. A. and Walker, S.: *Dynamic Analysis of Offshore Structures.* Newnes-Butterworth, London, 1979.

[12.8] The British Ship Research Association: *A Critical Evaluation of the Data on Wave Force Coefficients.* Report No. W 278, Aug. 1976.

[12.9] British Standards Institution: *Fixed Offshore Structures.* Draft for Development, DD55: 1978.

[12.10] Cartwright, D. E. and Longuet-Higgins, M. S.: *The Statistical Distribution of the Maxima of a Random Function.* Proc. Royal Society, A 237, 1956.

[12.11] Chakrabarti, S. K.: *Discussion on [12.25].* Journal Eng. Mech. Div., ASCE, EM3, June 1971.

[12.12] Davenport, A. G.: *Note on the Distribution of the Largest Value of a Random Function with Application to Gust Loading.* Proc. Institution of Civil Engineers, London, Vol. 28, 1964.

[12.13] Davenport, A. G.: *The Dependence of Wind Loads on Meteorological Parameters.* Proc. Int. Conf. on Wind Effects on Buildings and Structures, Ottawa, Sept. 1967, Vol. I.

[12.14] Draper, L. and Squire, E. M.: *Waves at Ocean Weather Ship Station India.* Trans. Royal Institute of Naval Architects, 1967.

[12.15] Fjeld, S., Andersen, T., Myklatun, B.: *Risk Analysis of Offshore Production and Drilling Platforms.* Proc. Offshore Tech. Conf. Paper 3152, Houston, 1978.

[12.16] Hallam, M. G., Heaf, N. J. and Wootton, L. R.: *Dynamics of Marine Structures.* Report UR8, CIRIA Underwater Engineering Group, 2nd edition, Aug. 1978.

[12.17] Hasselmann, K. et al.: *Measurements of Wind-wave Growth and Swell Decay during the Joint North Sea Wave Project (JONSWAP).* Deutsches Hydrographisches Zeitschrift, Hamburg, Reihe A (8°) Nr. 12.

[12.18] Hohenbichler, M. and Rackwitz, R.: *Non-Normal Vectors in Structural Reliability.* Journal Eng. Mech. Div. ASCE, EM6, Vol. 109, Dec. 1981.

[12.19] Houmb, O. G. and Overvik, T.: *Parameterization of Wave Spectra and Long Term Joint Distribution of Wave Height and Period.* Proc. 1st International Conference on the Behaviour of Offshore Structures, Trondheim, 1976.

[12.20] Kim, Y. Y. and Hibbard, H. C.: *Analysis of Simultaneous Wave Force and Water Particle Velocity Measurements.* Proc. Offshore Tech. Conf. Paper 2192, Houston, 1975.

[12.21] Kinsman, B.: *Wind Waves. Their Generation and Propagation on the Ocean Surface.* Prentice-Hall, Englewood Cliffs, N. J., 1965.

[12.22] Laboratorio Nacional de Engenharia Civil: *Wind in Western Europe. Results of an Enquiry.* Proc. 34/13/4047, LNEC, Lisbon, Feb. 1973.

[12.23] L. E. A. Offshore Management: *Permanently Located Offshore Structures - Jacket Sensitivity Study.* 1979 (unpublished).

[12.24] Malhotra, A. K. and Penzien, J.: *Response of Offshore Structures to Random Wave Loads.* Journal Str. Div. ASCE, ST10, Oct. 1970.

[12.25] Malhotra, A. K. and Penzien, J.: *Nondeterministic Analysis of Offshore Structures.* Journal Eng. Mech. Div. ASCE, EM6, Dec. 1970.

[12.26] Marshall, P. W. and Bea, R. G.: *Failure Modes of Offshore Platforms.* Proc. 1st International Conference on the Behaviour of Offshore Structures, Vol. 2, Trondheim, 1976.

[12.27] Moan, T. and Holand, I.: *Risk Assessment of Fixed Offshore Structures - Experience and Principles.* Proc. 3rd International Conference of Structural Safety and Reliability, Trondheim, 1981.

[12.28] Morison, J. R., O'Brien, M. P., Johnson, J. W. and Schoaf, S. A.: *The Forces Exerted by Surface Waves on Piles.* Petroleum Transactions, AIME, Vol. 189, 1950.

[12.29] Moses, F.: *Reliability Analysis Format for Offshore Structures.* Proc. Offshore Tech. Conf. Paper 3046, Houston, 1978.

[12.30] Newland, D. E.: *An Introduction to Random Vibrations and Spectral Analysis.* Longman, London, 1975.

[12.31] Pierson, W. J. and Moskowitz, L.: *A Proposed Spectral Form for Fully Developed Wind Seas based on the Similarity Theory of S. A. Kitaigordskii.* Journal of Geophysical Research, Vol. 69, No. 24, Dec. 1964.

[12.32] Rice, S. O.: *Mathematical Analysis of Random Noise.* Bell Systems Tech. Journal, Vol. 23, 1944. Reprinted in Wax, N. (ed.): *Selected Papers on Noise and Stochastic Processes.* Dover, N. Y., 1954.

[12.33] Rice, S. O.: *Mathematical Analysis of Random Noise.* Bell Systems Tech. Journal, Vol. 44, 1945 (Reprinted with [12.32]).

[12.34] Rosenblatt, M.: *Remarks on a Multi-Variate Transformation.* Annals of Mathematical Statistics, Vol. 23, 1952.

[12.35] Sarpkaya, T.: *The Hydrodynamic Resistance of Roughened Cylinders in Harmonic Flow.* Royal INst. Naval Architects, Spring Meeting, 1977, Paper No. 4.

[12.36] Stuugroep Problematiek von Offshore Constructies (Stu POC): *Probabilistic Reliability Analysis for Offshore Structures.* StuPOC-V-5-6. Netherlands Industrial Council for Oceanology. English Edition, 1980.

[12.37] Van der Hoven, I.: *Power Spectrum of Horizontal Wind Speed in the Frequency Range from 0.0007 to 900 Cycles per Hour.* Journal of Meteorology, Vol. 14, 1957.

[12.38] Wiegel, R. L.: *Waves and Wave Spectra and Design Estimates.* Conf. Deep-Sea Oil Production Structures. University of California, Berkeley, 1978.

Chapter 13

RELIABILITY THEORY AND QUALITY ASSURANCE

13.1 INTRODUCTION

The first 12 chapters of this book are devoted to various aspects of structural reliability theory and its application to design and safety checking. In undertaking a reliability analysis, the engineer should take account of all known sources of uncertainty and should use this information to control the probabilities of structural failure and unserviceability within acceptable ranges. This can be done either directly, by modifying some part of the structure, or indirectly by modifying the partial coefficients. Allowance should be made for the possibility of the occurrence of all recognised failure modes, e.g. shear, buckling, plastic collapse, etc., together with various modes of unserviceability.

It is widely recognised, however, that most structural failures occur for unexpected reasons and in ways that have not previously been encountered [13.2], [13.7], [13.11], [13.14], [13.17]. No discussion of structural reliability theory is therefore complete without some consideration being given to these additional causes of failure and their possible treatment.

13.2 GROSS ERRORS

13.2.1 General

In chapters 3 and 12 a considerable amount of space is used to discuss the probabilistic modelling of loads and resistance variables. These models are selected or devised in such a way as to embody those features of the physical quantity that are essential for the analysis of the practical problems being considered. It should not be thought that the models are intended to be a perfect mirror image of reality, but rather as a »tool» in a decision making process. Depending on the nature of the decision to be taken, the »tool» may need to be changed. For example, in modelling the yield stress of steel, a simple log-normal distribution may often be used (see chapter 3), but on other occasions a mixed distribution model (see equation (3.34)) would be more appropriate.

One important assumption that has been made is that the probabilistic models for loads and resistance variables are representative of events during a particular period of time - the life of the structure in the case of loads and other actions, and the period of construction in the case of material properties. In fact, the models are conditional upon or pre-suppose certain standards of design checking, quality control, inspection and maintenance. If, for example, the standards of quality control used in the manufacture of a structural material change, one would expect to see some change in the probability distribution function of that variable. If the standards of quality control are significantly different between different manufacturers or suppliers (e.g. in the case of steel or concrete), it may be convenient to use a mixed distribution model to allow for these differences.

The problems that need to be considered here, however, are of a different nature. The vast majority of structural failures occur because of *gross errors.* A *gross error* is defined as a major or fundamental mistake in some aspect of the processes of planning, design, analysis, construction, use or maintenance of a structure that has the potential for causing failure. Gross errors occur because of inadequacies in the standards of *quality assurance* - the process by which the various components of complete »building process» (mentioned above) are co-ordinated with the aim of achieving the design objective. It should be noted, however, that not all gross errors make a structure weaker - they can also make it unnecessary strong. Typical examples of gross errors are mistakes in design calculations, use of the wrong size of reinforcing bars or grade of steel, misinterpretations of geotechnical survey data, subjecting the structure to a class of loading for which it was not intended, etc.

A gross error should not therefore be considered as some extreme value in the tail of the probability distribution used to model a particular random variable, but a discrete event G which radically alters the probability of failure by changing the models that are applicable.

Example 13.1. Assume, for the sake of simplicity, that a structure has a resistance R which is dependent on only one basic variable, the yield stress of steel, and that it is subjected to a single load effect S. For simplicity, let R and S be normally distributed and

$$\mu_R = 380 \text{ N/mm}^2 \quad ; \quad \sigma_R = 25 \text{ N/mm}^2$$

$$\mu_S = 230 \text{ N/mm}^2 \quad ; \quad \sigma_S = 43 \text{ N/mm}^2$$

Then, from equations (4.26) and (4.27), $\beta = 3.02$ and $P_f = 1.3 \times 10^{-3}$.

Assume now that a lower grade of steel is used in place of the correct grade, so that R is distributed with parameters

$$\mu'_R = 280 \text{ N/mm}^2 \quad ; \quad \sigma'_R = 25 \text{ N/mm}^2$$

giving $\beta' = 1.0$ and $P'_f = 0.16$.

The values of μ_R and μ'_R given above are typical for the yield stress of grade 50 and 43 weldable structural steels, and it can therefore be seen that a gross error involving the substitution of grade 43 steel for grade 50 in a critical part of a structure is quite likely to cause failure - a chance of about 1 in 6 with the arbitrary assumptions made here.

The preceding example is somewhat simplistic and little attention should be paid to the actual numbers used. However, it illustrates the principle that the models which are used under normal conditions, without the presence of gross errors, are no longer applicable when a gross error occurs. This does not mean that reliability theory cannot be used under these conditions - it means that the *models* have to be amended. The problem is to know the various forms which gross errors can take.

13.2.2 Classification of gross errors

Table 13.1 gives a general classification of the nature and sources of gross errors, along with some examples. No such list can, of course, be comprehensive. Indeed, by their very nature, some potential gross errors or hazardous situations must exist which have not yet been recognised. Ignorance of phenomena such as fatigue, brittle fracture and the deterioration of concrete made from high-alumina cement are typical examples from the past of errors in design concept. Early designers cannot be criticised for not knowing about such effects, just as there is no reason to suppose that currently unrecognised failure modes will not cause accidents in the future.

Those gross errors which cause structural collapse or unserviceability can also be classified according to the type of failure which occurs, as shown in table 13.2. The word failure, here, should be interpreted in the general sense of failure to comply with some performance requirement and not just collapse (i.e. it includes unserviceability). Two types are given: those in which the structure fails in a predictable manner by one of a number of foreseen failure modes - here called type A; and those in which unforeseen failure modes occur - called type B.

Source	Nature	Example
Design concept	- Possible failure mode unrecognised	... neglect of lateral torsional buckling
	- Incorrect nature of use assumed	... room used for storage of heavy equipment in office premises
Design and analysis	- Omission of a load or load combination	... effect of ground-water pressure overlooked
	- Misinterpretation of geotechnical data	... soft stratum not detected
	- Computational error in analysis	... error in computer program
	- Misinterpretation of units	... kilogrammes interpreted as Newtons
	- Error in detailing	... 20 mm bars used instead of 40 mm
Construction	- Misinterpretation of drawings	... 100 mm slab instead of 150 mm
	- Use of incorrect material	... grade 43 steel used instead of grade 50
	- Incorrect fabrication	... omission of heat treatment
	- Incorrect construction	... error in position of reinforcement
Inspection	- Gross defect not detected	... crack in weld
Use	- Accidental loading	... severe impact or explosion
	- Change of use without structural assessment	... domestic premises used for public library
	- Need for specialist maintenance overlooked	... cathodic protection system becomes inoperative

Table 13.1. General classification of the nature and sources of gross errors.

Type of failure	Gross error
Type A: Failure in a mode of behaviour against which the structure was designed	*Errors affecting:* - load-carrying capacity - ability to remain serviceable - applied load(s)
Type B: Failure in a mode of behaviour against which the structure was not designed	*Errors that relate to the fundamental understanding of structural behaviour, arising from:* - profession's ignorance - engineers' ignorance - engineers' oversight

Table 13.2. Classification of gross errors according to type of failure.

Failures of type A arise because of gross errors in the calculation of design loads and/or load-carrying capacities, and/or because of weaknesses which are introduced into the structure during construction. Most of the examples in table 13.1 are of this type. Failures of type B occur mainly because of lack of knowledge. However, distinction must be made between failures which are the result of ignorance within the structural engineering profession as a whole and those that occur because of ignorance or negligence by an individual or design team. In the latter case, the engineer or team is clearly responsible for the failure. In the former, he or they are just unfortunate, unless it can be shown that currently accepted practice has been extrapolated to an unreasonable extent.

The prevention of failures which arise from lack of knowledge within the profession as a whole is clearly impossible and occasional failures of this type will continue to occur. They will then be researched and this will add to the general fund of engineering knowledge. Failures of type A, and failures of type B resulting from an *engineer's* ignorance or negligence, are in theory preventable, but this requires an appropriate level of expenditure on education, training, design checking, quality control, inspection, maintenance, etc. The planning and co-ordination of these various tasks is the subject of *quality assurance.* This is briefly referred to in section 13.4.

Tables 13.1 and 13.2 show only two of many possible ways of classifying gross errors. For instance, errors can be classified according to

- nature of the error (table 13.1)
- type of failure associated with the error (table 13.2)
- consequences of failure arising from the error
- those responsible for causing the error
- those responsible for not detecting the error
- etc.

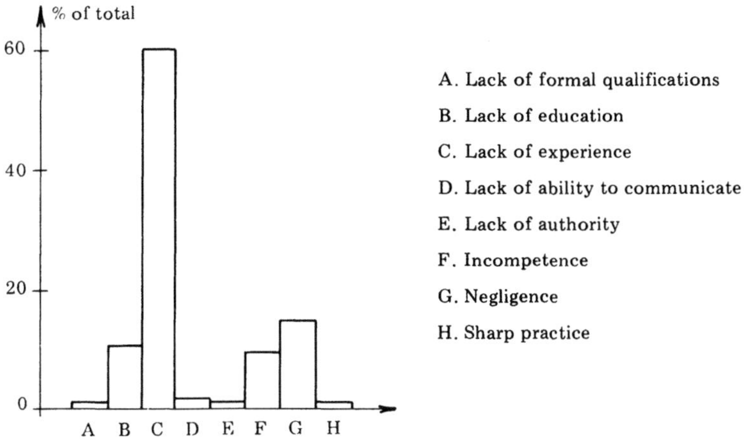

Figure 13.1. Analysis of underlying causes of 120 structural failures in buildings, from [13.8].

Considerable success has been achieved in the analysis of structural failure data using this type of classification [13.8], [13.11], [13.17], [13.18]. An important conclusion is that many gross errors occur because of lack of experience on the part of those undertaking the work and because the fundamental behaviour of the structure is often not fully understood. Figure 13.1, taken from [13.8], illustrates this point. In a study of 120 building failures, it was found that over 60% were due to lack of experience by the personnel concerned and that in about 50% of all cases the major cause of failure was a lack of appreciation of the relevant design concepts - for example, ignorance of the need to design against lateral torsional buckling in unsupported compression flanges.

Such findings are of considerable value in the planning of quality assurance schemes, but this will not be discussed here.

13.3 INTERACTION OF RELIABILITY THEORY AND QUALITY ASSURANCE

13.3.1 General

The phrase »building process» has been used in the preceding sections as a general term to include planning, design, analysis, construction, maintenance and use of a structure. It will also be used here, to cover all aspects and stages of a structural development, but it should be emphasised that it is applicable to all structural projects, not just buildings.

For a given structure and location, the building process can be divided into two distinct stages:

1) preparing a precise deterministic specification for the structure, and
2) building the structure and checking that the specification is met.

The specification, involving documents and drawings, will typically contain information of the type

- the column shall contain 16 40 mm diameter reinforcing bars
- the nominal thickness of the slab shall be 200 mm
- the structural steel shall have a nominal yield stress of 350 N/mm^2
- etc

This information, which is passed from the designer to the contractor (or builder), is and needs to be of a deterministic nature. The second part of the building process involves the transformation of the specification into physical reality - the structure - and checking that it is satisfactory.

For new structures, the role played by reliability theory is in the preparation of the structural specification, either directly, by subjecting the proposed design to a reliability analysis or, indirectly, by using a code in which the partial coefficients have been assessed probabilistically. As previously mentioned (for example in chapter 3), the probabilistic models used in reliability-analysis are conditional upon the specified standards of quality control and acceptance tests for the materials, and on the standards of inspection for the finished structure. Stages 1) and 2) of the building process are (or should be) intimately related. The role of quality assurance, in its broadest sense, is to coordinate, rationalise and monitor these two stages.

Reliability theory also has a role to play in the assessment of existing structures, particularly when structural damage has occurred as a result of accidental loading, or when a structure is being assessed for a radical change of use. It should be noted that in all cases, the questions to be answered are of the type: Is the structure strong enough? Should the nominal dimensions be increased or reduced, and if so by how much?

13.3.2 The effect of gross errors on the choice of partial coefficients

The problem that must now be considered is whether knowledge that gross errors can occur during the processes of design and construction should affect the rational choice of partial coefficients for use in level 1 codes. This is best explored by means of an example.

Example 13.2. (taken from [13.1]). Take a failure function of the form f(G, R, K, S) = 0 and let the safety margin M be

$$M = GR - KS \tag{13.1}$$

where

R is a continuously distributed random strength variable, $N(\mu_R, \sigma_R)$

S is a continuously distributed random load effect, $N(\mu_S, \sigma_S)$

K is a discrete-valued model uncertainty, and

G is a gross error which modifies the strength parameter R,

and assume that the quantities R, S, K and G are statistically independent. This is a reasonable assumption, as least as far as independence of G and the other variables is concerned, since it may be assumed, for example, that the probability of having an incorrect size of reinforcing bars is unrelated to the yield stress of the bars or to the loads that are subsequently applied to the structure.

Let the initial cost of the structure be given by

$$c_I = \alpha \mu_R \quad (13.2)$$

where α is a constant, and let the consequential cost of failure, should it occur, be

$$c_f = \xi c_I = \xi \alpha \mu_R \quad (13.3)$$

The probability of failure, given a gross error of magnitude g and a model uncertainty of magnitude k, may be expressed as

$$P_f | g, k = P(gR - kS \leqslant 0) \quad (13.4)$$

and the expected conditional total cost as

$$E[C_T | g, k] = c_I + c_f P_f | g, k \quad (13.5)$$

Assuming that the model uncertainty K can take the following discrete values with probability mass $p(k_i)$

k_i	0.4	0.5	0.6
$p(k_i)$	0.2	0.6	0.2

Table 13.3

the expected total cost, given a gross error g, $E[C_T | g]$, is

$$E[C_T | g] = \sum_{i=1}^{3} (c_I + c_f P_f | g) p(k) \quad (13.6)$$

The expected total cost given that there is a gross error g with a probability p(g), and no error (g = 1) with a probability 1 − p(g) is thus

$$E[C_T(g)] = \sum_{i=1}^{3} (c_I + c_f P_f | g) p(k) p(g) + \sum_{i=1}^{3} (c_I + c_f P_f) p(k)(1 - p(g)) \quad (13.7)$$

Substituting for c_I and c_f from equations (13.2) and (13.3) gives

$$E[C_T(g)] = \sum_{i=1}^{3} \alpha \mu_R (1 + \xi P_f | g) p(k) p(g) + \sum_{i=1}^{3} \alpha \mu_R (1 + \xi P_f) p(k)(1 - p(g)) \quad (13.8)$$

Variable	μ	σ/μ
R	-	0.07
S	100	0.10
α = 10 cost units		
ξ = 20		

Table 13.4

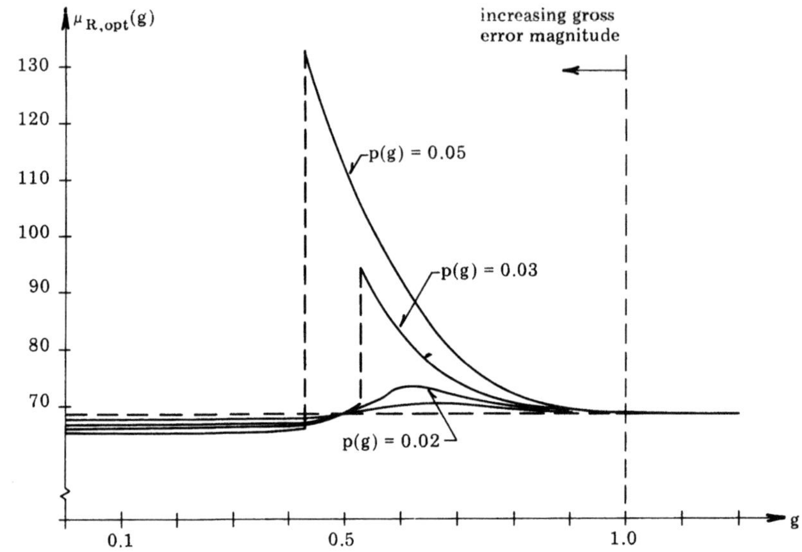

Figure 13.2

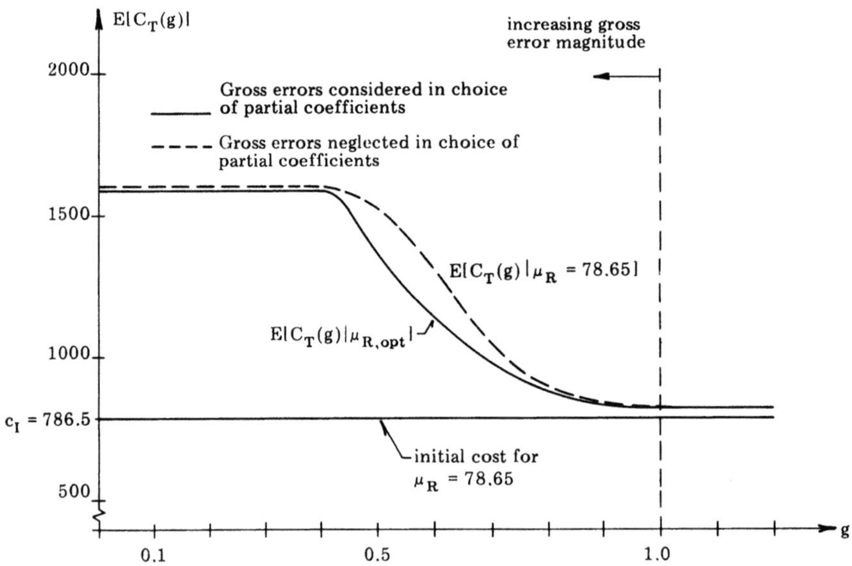

Figure 13.3 ($p(g) = 0.05$).

: Let us now undertake an unconstrained minimisation of $E[C_T(g)]$ with respect to the
: quantity μ_R and denote the minimum value of μ_R by $\mu_{R,opt}(g)$. This is shown in figure
: 13.2 for the set of parameters given in table 13.4. An important result for this particular
: set of assumptions is that if the quality assurance process is able to restrict the frequency
: of gross errors to less than 2%, then $\mu_{R,opt}$ is very insensitive to the occurrence of gross
: errors of any magnitude. A more important result, however, is that even if the frequency
: of gross errors rises to say 5%, although, $\mu_{R,opt}$ shows a marked increase for gross errors
: of moderate magnitude, the expected total cost given by

$$E[C_T(g)|\mu_{R,opt}] = \sum_{i=1}^{3}(c_I + c_f P_f|g, \mu_{R,opt})p(k)p(g)$$

$$+ \sum_{i=1}^{3}(c_I + c_f P_f|\mu_{R,opt})p(k)(1 - p(g)) \qquad (13.9)$$

: is insensitive to the decision of whether or not to allow for the possibility of gross errors
: in calculating $\mu_{R,opt}$ (see figure 13.3). At the worst, the total expected cost differs by
: only 15%.

This example indicates that, at least for the set of models and parameters chosen, the possibility of the occurrence of gross errors should not influence the selection of partial coefficients for use in structural design. Some further results are given in [13.1]. The extent to which these results can be generalised depends on circumstances, but it is considered that under many conditions the optimisation of expenditure on the control of gross errors can be undertaken independently of the choice of partial coefficients. Similar conclusions have been reached by others, e.g. [13.6].

13.4 QUALITY ASSURANCE

The respective values of reliability analysis and quality assurance have been explored earlier in this chapter and have been shown to be entirely compatible. The analysis of many structural failures (see e.g. [13.2]) shows that the majority could not have been prevented by minor increases in partial coefficients. This is consistent with the results obtained from example 13.2 and indicates that relatively more resources should be deployed on control, inspection and checking - i.e. quality assurance. A discussion of this large subject is beyond the scope of the present text but the reader is referred to [13.9] for further study.

BIBLIOGRAPHY

[13.1] Baker, M. J. and Wyatt, T. A.: *Methods of Reliability Analysis for Jacket Platforms.* Proc. 2nd International Conference on the Behaviour of Off-Shore Structures, London, 1979.

[13.2] Blockley, D. I.: *Analysis of Structural Failures.* Proc. Institution of Civil Engineers, Part 1, Vol. 62, Feb. 1977.

[13.3] Blockley, D. I.: *The Nature of Structural Design and Safety*. Ellis Horwood, Chichester, 1980.

[13.4] Bosshard, W.: *Structural Safety - A Matter of Decision and Control*, IABSE Surveys S 9/79, IABSE Periodica 2/1979.

[13.5] CIRIA: *Rationalisation of Safety and Serviceability Factors in Structural Codes*. Construction Industry Research and Information Association. Report No. 63, 1977.

[13.6] Ditlevsen, O.: *Formal and Real Structural Safety. Influence of Gross Errors*. Lectures on Structural Reliability (ed. P. Thoft-Christensen), Aalborg University Centre, Aalborg, Denmark, 1980, pp. 121-147.

[13.7] Feld, J.: *Lessons from Failures of Concrete Structures*. ACI Monograph No. 1, 1964.

[13.8] Institution of Structural Engineers: *Structural Failures in Buildings*. Symposium, London, April 1980.

[13.9] Joint Committee on Structural Safety, CEB - CECM - CIB - FIP - IABSE - IASS - RILEM: *General Principles on Quality Assurance for Structures*. IABSE. Reports of the Working Commissions. Vol. 35, 1981.

[13.10] Matousek, M.: *Massnahmen gegen Fehler im Bauprozess*. Dissertation No. 6941. Eidgenössische Technische Hochschule (ETH), Zürich, 1981.

[13.11] Matousek, M. and Schneider, J.: *Untersuchungen zur Struktur des Sicherheitsproblems bei Bauwerken*. Bericht Nr. 59, Institut für Baustatik und Konstruktion, ETH, Zürich, 1976.

[13.12] Melchers, R. E.: *Selection of Control Levels for Maximum Utility of Structures*. Proc. 3rd International Conference on the Application of Statistics and Probability in Soil and Structural Engineering. Sydney, 1979.

[13.13] Moan, T. and Holand, I.: *Risk Assessment of Offshore Structures - Experience and Principles*. Proc. 3rd International Conference on Structural Safety and Reliability, Trondheim, 1981.

[13.14] Pugsley, A. G.: *The Safety of Structures*. Arnold, London, 1966.

[13.15] Schneider, J.: *Organisation and Management of Structural Safety during Design, Construction and Operation of Structures*. Proc. 3rd International Conference on Structural Safety and Reliability, Trondheim, 1981.

[13.16] SIA: *Weisung für die Koordination des Normwerks des SIA im Hinblick auf Sicherheit und Gebrauchsfähigkeit von Tragwerken*. SIA 260, Schweiz. Ingenieur- und Architekten-Verein (SIA), Zürich, 1981.

[13.17] Sibly, P. G. and Walker, A. C.: *Structural Accidents and their Causes*. Proc. Institution of Civil Engineers, Part 1, Vol. 82, May 1977.

[13.18] Tveit, O. J. and Evandt, Ø.: *Experiences with Failures and Accidents of Offshore Structures*. Proc. 3rd International Conference on Structural Safety and Reliability, Trondheim, 1981.

Appendix A
RANDOM NUMBER GENERATORS

1. GENERAL

For simulation studies and Monte-Carlo analysis use has to be made of long sequences of random numbers (generally pseudo-random numbers). These are most conveniently generated using a digital computer. The increased use of such studies in recent years has meant that reliable library functions have been made available on most computer systems. It can normally be assumed that these functions have been fully tested for random behaviour.

2. UNIFORM RANDOM NUMBER GENERATORS

Most digital random number generators are based on uniform pseudo-random number generators of the *multiplicative congruence type.* A uniform random number generator is one which generates successive independent realisations u_i of a random variable U having a rectangular density function, usually in the interval [0, 1], i.e.

$$f_U(u) = \begin{cases} 1 & 0 \leqslant u \leqslant 1 \\ 0 & \text{elsewhere} \end{cases} \tag{A.1}$$

giving

$$F_U(u) = \begin{cases} 0 & u < 0 \\ u & 0 \leqslant u \leqslant 1 \\ 1 & u > 1 \end{cases} \tag{A.2}$$

The mean and standard deviation of the random variable U can be shown to be

$$\mu_U = 0.5 \quad , \quad \sigma_U = 1/12$$

3. MULTIPLICATIVE CONGRUENCE METHOD

This method produces a series of pseudo-random numbers r_i that eventually repeats, but, if correctly designed, only after a very long cycle. The next number in the pseudo-random series is related to the previous number by the relationship

$$r_{n+1} = a r_n \text{ (modulo m)} \tag{A.3}$$

where a and m are integer constants and are relatively prime. Starting with an integer »seed» i_0 the first pseudo-random number r_1 in the interval [0, 1] is obtained from

$$\frac{a i_0}{m} = j_1 + \frac{a i_0 - j_1 m}{m} = j_1 + \frac{i_1}{m} = j_1 + r_1 \tag{A.4}$$

i.e. r_1 is the fractional part of the quotient $(a i_0)/m$ and j_1 is the integer part.

$i_1 = (a i_0 - j_1 m)$ is the seed for the second random number.

In general, the integer constants a and m are chosen to obtain the longest possible cycle. It can be shown that if

$$m = 2^b$$

and

$$a = (8t \pm 3)$$

where b and t are integers, then the length of the integer sequence before repetition is of the order of $2^{(b-2)}$.

Example A.1. Let $a = 5^5 = (8 \times 391 - 3) = 3125$ and $m = 2^{26} = 67108864$. Let the starting seed $i_0 = 1234567$ ($< 2^{26}$).

Then

$$\frac{a i_0}{m} = \frac{3125 \times 1234567}{67108864} = j_1 + \frac{i_1}{m} = 57 + \frac{32816627}{67108864} = 57.48901$$

The first random number is the fractional part of $a i_0/m$ = 0.48901 = r_1 and the new seed for generating the second number is i_1 = 32816627. The first few terms of this random sequence are:

Term n	i_n	r_n
0	1234567	-
1	32816627	0.48901
2	9615183	0.14328
3	49784667	0.74185
		0.27921

The length of this sequence before repetition is approximately $2^{(26-2)} = 2^{24} = 16777216$.

A commonly used random number generator on, for example the CDC 6000 Series computers is

$$r_{n+1} = 186277\, r_n \quad (\text{modulo } 2^{48})$$

This has approximately $2^{46} \simeq (7.04 \times 10^{13})$ random numbers before repetition, a sufficiently large number for most purposes.

The pseudo-random numbers r_i generated by this method can be assumed to be independent realisations u_i of a random variable U having a rectangular distribution with $0 \leqslant U \leqslant 1$.

4. GENERATION OF RANDOM DEVIATES WITH A SPECIFIED PROBABILITY DISTRIBUTION FUNCTION F_X

A convenient general method consists of generating a random number, r, as described above and then, by making use of equation (A.2), finding the corresponding random deviate x of the random variable X from

$$F_X(x) = F_U(u) = u = r \tag{A.5}$$

where F_X is the required distribution function. It is therefore necessary to find the inverse function F^{-1}, giving

$$x = F_X^{-1}(r) \tag{A.6}$$

This is valid for all forms of distribution function, but two classes of variable exist which require different treatment.

Class A: The distribution function F_X has an inverse F_X^{-1} which can be expressed in closed form

In this case the random deviate x can be generated simply by obtaining successive values

$$x_i = F_X^{-1}(r_i) \tag{A.7}$$

Example A.2. Let X be type I extreme (maxima) distributed with distribution function

$$F_X(x) = \exp[-\exp(-\alpha(x-\gamma))] \tag{A.8}$$

and parameters γ and α. Then

$$x = \gamma - \frac{1}{\alpha} \ln(-\ln F_X(x)) \tag{A.9}$$

: Hence, a sequence of independent random deviates x_i of the random variable X may be
: obtained from
:
: $$x_i = \gamma - \frac{1}{\alpha} \ell n(-\ell n r_i) \tag{A.10}$$
:
: where r_i are rectangularly distributed random numbers in the interval [0, 1].

Class B: The distribution function F_X has an inverse F_X^{-1} which cannot be expressed in closed form

In this case the general procedure is the same, but the inverse function has to be evaluated either graphically, by numerical integration, by table look-up and interpolation, or by fitting an appropriate polynomial. The last three methods are suitable for computer application.

5. SPECIAL CASES: GENERATION OF RANDOM DEVIATES HAVING NORMAL AND LOG-NORMAL DISTRIBUTIONS

The normal and log-normal distributions are two of the Class B functions, but because of their frequent use they deserve further attention. In addition to the general method described above, a number of special methods exist for normal variables. These methods may also be used for generating log-normally distributed random numbers, by the use of an appropriate transformation.

Generation of random normal deviates from the sum of n rectangularly distributed random deviates.

The fact that under very general conditions, the distribution function for the sum of a series of independent random variables tends to a normal distribution as the number of variables in the sum increases (refer to the central limit theorem in chapter 3) can be used to generate random numbers having a distribution which approximates very closely to normal.

Most computer routines use the sum of 12 or more independent rectangularly distributed random numbers r_i. If the latter are generated in the interval [0, a], their sum ξ can easily be shown to be approximately normally distributed with mean μ_ξ given by

$$\mu_\xi = an/2 \tag{A.11}$$

and variance σ_ξ^2 by

$$\sigma_\xi^2 = a^2 n/12 \tag{A.12}$$

For the simple case when a = 1 and n = 12, ξ' given by

$$\xi' = \sum_{i=1}^{12} r_i - 6 \tag{A.13}$$

is approximately normally distributed with zero mean and unit standard deviation.

This approach gives excellent approximations to the normal distribution for deviates within two or three standard deviations from the mean, but for extreme values the approximation becomes increasingly poor, unless n is large. For example, the random variable ξ' defined by equation (A.13) cannot lie outside the the interval $[-6, 6]$.

Generation of random normal deviates using method due to Box and Muller

Box and Muller [A.1] have shown that if r_1 and r_2 are *independent* random variables from the same rectangular distribution in the interval $[0, 1]$, then N_1 and N_2 given by

$$\begin{aligned} N_1 &= (-2\,\ell n\, r_1)^{\frac{1}{2}} \cos 2\pi r_2 \\ N_2 &= (-2\,\ell n\, r_1)^{\frac{1}{2}} \sin 2\pi r_2 \end{aligned} \qquad \text{(A.14)}$$

are independent random variables, normally distributed with zero mean and unit variance.

The advantage of this method is that it is accurate over the complete range and depends only on the randomness and independence of r_1 and r_2.

BIBLIOGRAPHY

[A.1] Box, G. E. P. and Muller, M. E.: *A Note on the Generation of Random Normal Deviates.* Annals of Math. Statistics, Vol. 29, 1958.

[A.2] Hammersley, J. M. and Handscombe, D. C.: *Monte Carlo Methods.* Methuen, 1964.

[A.3] Tocher, K. D.: *The Art of Simulation.* The English Universities Press, 1963.

Appendix B
SPECTRAL ANALYSIS OF WAVE FORCES

1. INTRODUCTION

The purpose of this appendix is to derive a relationship between the spectrum of water surface elevation $S_{\eta\eta}(\omega)$ and the spectrum of structural displacements $S_{SS}(\omega)$ for a typical multipile jacket structure in a given normal response mode. The theory is derived from the work of Borgman [B.1] and Malhotra and Penzien [B.3]. For a discussion of the general theory of random vibrations and spectral analysis see, for example, [B.2] or [B.4].

2. GENERAL EQUATIONS OF MOTION

Assuming that the structure can be idealised as a lumped mass system (see figure B.1), the equations of motion may be expressed in the well-known form

$$\bar{\bar{m}}\ddot{\bar{u}}_s + \bar{\bar{c}}_s\dot{\bar{u}}_s + \bar{\bar{k}}\bar{u}_s = \bar{p}(t) \qquad (B.1)$$

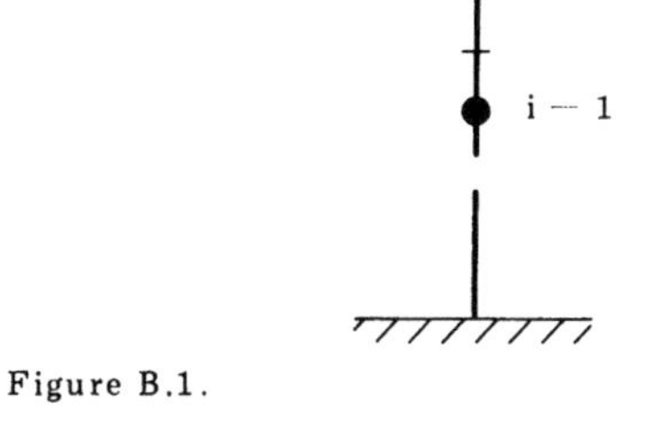

Figure B.1.

where

$\bar{\bar{m}}$ is a diagonal mass matrix

$\bar{\bar{c}}_s$ is a structural damping matrix

$\bar{\bar{k}}$ is a structural stiffness matrix

$\bar{u}_s$ is a vector of displacements of the various lumped masses

$\ddot{\bar{u}}_s = \frac{d\dot{\bar{u}}_s}{dt} = \frac{d^2\bar{u}_s}{dt^2}$

and where $\bar{p}(t)$ is a vector of wave loads $p_i(t)$ acting on the lumped masses

$$p_i(t) = \int_{(z_i + z_{i-1})/2}^{(z_i + z_{i+1})/2} dP(t) \tag{B.2}$$

Substituting for dP(t) from equation (12.37), equation (B.1) becomes

$$\bar{\bar{m}}\ddot{\bar{u}}_s + \bar{\bar{c}}_s\dot{\bar{u}}_s + \bar{\bar{k}}\bar{u}_s = \bar{\bar{c}}_a(\dot{\bar{u}}_p - \dot{\bar{u}}_s)|\dot{\bar{u}}_p - \dot{\bar{u}}_s| + \bar{\bar{c}}_b(\ddot{\bar{u}}_p - \ddot{\bar{u}}_s) + \rho\bar{\bar{A}}\ddot{\bar{u}}_s \tag{B.3}$$

where

$\dot{\bar{u}}_p$ is a vector of water particle velocities

$\bar{\bar{c}}_a$ is a matrix involving parameters such as C_d, D, ρ (see chapter 12)

$\bar{\bar{c}}_b$ is a matrix involving parameters such as C_m, ρ, A (see chapter 12)

$\bar{\bar{A}}$ is a diagonal matrix of cross-sectional areas.

Let $\bar{r} = \bar{u}_p - \bar{u}_s$. Equation (B.3) may now be written in terms of $\dot{\bar{r}}$ and $\ddot{\bar{r}}$ as

$$(\bar{\bar{m}} + \bar{\bar{c}}_b - \rho\bar{\bar{A}})\ddot{\bar{r}} + \bar{\bar{c}}\dot{\bar{r}} + \bar{\bar{k}}\bar{r} + \bar{\bar{e}} = (\bar{\bar{m}} - \rho\bar{\bar{A}})\ddot{\bar{u}}_p + \bar{\bar{c}}_s\dot{\bar{u}}_p + \bar{\bar{k}}\bar{u}_p \tag{B.4}$$

with

$$\bar{\bar{e}} = (\bar{\bar{c}}_s - \bar{\bar{c}})\dot{\bar{r}} + \bar{\bar{c}}_a\dot{\bar{r}}|\dot{\bar{r}}| \tag{B.5}$$

The right hand side of equation (B.4) is completely defined for a given wave frequency, containing terms involving only water particle displacement, velocity and acceleration. The terms on the left hand side are of the standard form for a linear multi-degree of freedom system with viscous damping, except for the error term $\bar{\bar{e}}$ which involves a quadratic term in $\dot{\bar{r}}$.

In order to make equation (B.4) linear, $\bar{\bar{e}}$ is minimised in the mean square sense and then dropped (see [B.3]). This leads to a damping matrix $\bar{\bar{c}}$ with terms

$$\left.\begin{aligned} c_{ii} &= c_{s,ii} + c_{a,ii}\sqrt{\frac{8}{\pi}}\,\sigma_{\dot{r},i} \\ c_{ij} &= 0 \quad , \quad i \neq j \end{aligned}\right\} \qquad (B.6)$$

where $\sigma_{\dot{r},i}$ is the square root of the relative velocity variance at the level of mass i. Hence

$$\bar{\bar{c}} = \bar{\bar{c}}_s + \bar{\bar{c}}_f \qquad (B.7)$$

where $\bar{\bar{c}}_f$ is a diagonal matrix of terms, referred to as fluid damping coefficients.

On re-arrangement, equation (B.4) then becomes

$$(\bar{\bar{m}} + \bar{\bar{c}}_b - \rho\bar{\bar{A}})\ddot{\bar{u}}_s + \bar{\bar{c}}\dot{\bar{u}}_s + \bar{\bar{k}}\bar{u}_s = \bar{\bar{c}}_f\dot{\bar{u}}_p + \bar{\bar{c}}_b\ddot{\bar{u}}_p \qquad (B.8)$$

The term $(\bar{\bar{m}} + \bar{\bar{c}}_b - \rho\bar{\bar{A}})$ is the effective mass matrix, $\bar{\bar{c}}$ is a damping matrix including components of structural and fluid damping, $\bar{\bar{c}}_f\dot{\bar{u}}_p$ is the contribution to the loading due to drag, and $\bar{\bar{c}}_b\ddot{\bar{u}}_p$ is the contribution to the loading due to inertia effects.

3. MODAL ANALYSIS

The matrix equation (B.8) represents a set of coupled differential equations, which can be uncoupled by using the system's eigenvector matrix (mode shape matrix), $\bar{\bar{\Psi}}$, as a transformation matrix. ($\bar{\bar{\Psi}}$ is obtained by ignoring the damping term $\bar{\bar{c}}\dot{\bar{u}}_s$).

Let

$$\bar{u}_s = \bar{\bar{\Psi}}\,\bar{x} \qquad (B.9)$$

Pre-multiplication by $\bar{\bar{\Psi}}^T$ and substitution in equation (B.8) gives

$$\bar{\bar{\Psi}}^T(\bar{\bar{m}} + \bar{\bar{c}}_b - \rho\bar{\bar{A}})\bar{\bar{\Psi}}\ddot{\bar{x}} + \bar{\bar{\Psi}}^T\bar{\bar{c}}\bar{\bar{\Psi}}\dot{\bar{x}} + \bar{\bar{\Psi}}^T\bar{\bar{k}}\bar{\bar{\Psi}}\bar{x} = \bar{\bar{\Psi}}^T\bar{\bar{c}}_f\dot{\bar{u}}_p + \bar{\bar{\Psi}}^T\bar{\bar{c}}_b\ddot{\bar{u}}_p \qquad (B.10)$$

or

$$\bar{\bar{M}}\ddot{\bar{x}} + \bar{\bar{C}}\dot{\bar{x}} + \bar{\bar{K}}\bar{x} = \bar{P}(t) \qquad (B.11)$$

The generalised mass matrix $\bar{\bar{M}}$ and stiffness matrix $\bar{\bar{K}}$ will be diagonal due to orthogonality but the damping matrix $\bar{\bar{C}}$ may be full. $\bar{\bar{C}}$ may be diagonalised by error minimization as in the drag term linearization. However, for highly damped systems, it is sufficiently accurate to use the diagonal terms of $\bar{\bar{C}}$ and to ignore the off-diagonal terms. This approach is followed below.

Equation (B.11) consists of a set of uncoupled modal equations which can be used to obtain the structural response in each »normal» mode separately, treating each as a one degree of freedom system.

4. SOLUTION STRATEGY

Consideration will be restricted to response in a single mode. Thus, equation (B.11) may be rewritten for mode j as ($M = M_{jj}$, $C = C_{jj}$, $K = K_{jj}$)

$$M\ddot{x}_j + C\dot{x}_j + Kx_j = P_j(t) = \bar{\Psi}_j^T(\bar{\bar{c}}_f \bar{\dot{u}}_p + \bar{\bar{c}}_b \bar{\ddot{u}}_p) \tag{B.12}$$

or

$$M\ddot{x}_j + C\dot{x}_j + Kx_j = \sum_{i=1}^{n} \Psi_{ji} P_i(t) \tag{B.13}$$

where n is the number of lumped masses and $P_i(t) = c_{f,ii}\,\dot{u}_{p,i} + c_{b,ii}\ddot{u}_{p,i}$.
Using complex number representation ($\iota = \sqrt{-1}$), let

$$x_j = a e^{\iota(\omega t + \phi)} \tag{B.14}$$

$$\dot{x}_j = \iota\omega a e^{\iota(\omega t + \phi)} = \iota\omega x_j \tag{B.15}$$

$$\ddot{x}_j = \iota^2 \omega^2 a e^{\iota(\omega t + \phi)} = -a\omega^2 x_j \tag{B.16}$$

Substitution for $\dot{x}_j$ and $\ddot{x}_j$ in equation (B.13) gives

$$(-\omega^2 M + \iota\omega C + K)x_j = \sum_{i=1}^{n} \Psi_{ji} P_i(t) \tag{B.17}$$

So

$$x_j(\omega) = \frac{\sum_{i=1}^{n} \Psi_{ji} P_i(t)}{(K-\omega^2 M)+\iota\omega C} = H(\iota\omega) \sum_{i=1}^{n} \Psi_{ji} P_i(t) \tag{B.18}$$

where $H(\iota\omega)$ is known as the complex frequency response function.
Thus

$$\dot{x}_j(\omega) = \iota\omega x_j = \iota\omega H(\iota\omega) \sum_{i=1}^{n} \Psi_{ji} P_i(t) \tag{B.19}$$

and the structural velocity for each mass i responding in mode j is given from equation (B.9) as

$$\dot{u}_{s,ij}(\omega) = \Psi_{ij} x_j(\omega) = \Psi_{ij} \iota\omega H(\iota\omega) \sum_{i=1}^{n} \Psi_{ji} P_i(t) \tag{B.20}$$

It can be seen from equations (B.3) to (B.8) that the structural displacement for any given component of wave frequency ω can be evaluated only if the linearised damping coefficients are known. Unfortunately, the damping coefficients are dependent on the variance of the relative velocity between the structure and the water particles (see equation (B.6)) requiring an iterative form of solution as described below.

Particle velocity variance: For a given component frequency and depth z, measured from the mean water level (positive z measured upwards), the particle velocity is obtained from linear wave theory (equations (12.29) and (12.31)) as

$$\dot{u}_p(z, \eta, \omega) = \omega \frac{\cosh k(z + d)}{\sinh kd} \eta \tag{B.21}$$

The corresponding water particle velocity spectrum $S_{\dot{u}_p \dot{u}_p}(z, \omega)$ is given by

$$S_{\dot{u}_p \dot{u}_p}(z, \omega) = \dot{u}_p(\omega) \dot{u}_p^{\star}(\omega) \tag{B.22}$$

where $\dot{u}_p^{\star}(\omega)$ is the complex conjugate of $\dot{u}_p$. Thus,

$$S_{\dot{u}_p \dot{u}_p}(z, \omega) = \omega^2 \frac{\cosh^2 k(z + d)}{\sinh^2 kd} S_{\eta\eta}(\omega) \tag{B.23}$$

and finally the particle velocity variance $\sigma^2_{\dot{u}_p}(z)$ by

$$\sigma^2_{\dot{u}_p}(z) = \int_0^{\infty} S_{\dot{u}_p}(z, \omega) d\omega \tag{B.24}$$

Force acting on a single pile: The generalised force, $P_j(t)$, obtained by taking the set of terms corresponding to mode j from the right hand side of equation (B.10), is given by

$$P_j(t) = \sum_{i=1}^{n} \Psi_{ji} P_i(t) = \sum_{i=1}^{n} \Psi_{ji}(c_{f,ii} \dot{u}_{p,i} + c_{b,ii} \ddot{u}_{p,i}) \tag{B.25}$$

Substituting for $c_{f,ii}$ and $\dot{u}_{p,i}$ from equations (B.6) and (B.21) and noting that by combining equations (12.29) and (12.32)

$$\ddot{u}_p(z, \eta, \omega) = \iota\omega^2 \frac{\cosh k(z + d)}{\sinh kd} \eta \tag{B.26}$$

equation (B.25) becomes

$$P_j(\omega) = \sum_{i=1}^{n} \Psi_{ji}(c_{a,ii} \sqrt{\frac{8}{\pi}} \sigma_{\dot{r}(i)} \omega \frac{\cosh k(z_i + d)}{\sinh kd} + \iota c_{b,ii} \omega^2 \frac{\cosh k(z_i + d)}{\sinh kd}) \eta \tag{B.27}$$

The only term in the above which is not known is $\sigma_{\dot{r},i}$, the square root of the relative velocity variance. At the commencement of the iterative calculation, it may be assumed that $\sigma_{\dot{r},i}$ equals $\sigma_{\dot{u}_p,i}$, which is given by equation (B.24).

Structural velocity variance: The structural velocity $\dot{u}_{s,ij}(\omega)$ at the level of the ith mass is given by equation (B.20) as

$$\dot{u}_{s,ij}(\omega) = \Psi_{ij}\iota\omega H(\iota\omega)\sum_{i=1}^{n}\Psi_{ji}P_i(t) = \Psi_{ij}\iota\omega H(\iota\omega)P_j(\omega) \tag{B.28}$$

where $P_j(\omega)$ is defined by equation (B.27).

The structural velocity spectrum $S_{ss,ij}(\omega)$ is obtained from

$$S_{ss,ij}(\omega) = \dot{u}_{s,ij}(\omega)\dot{u}^{\star}_{s,ij}(\omega) \tag{B.29}$$

where $\dot{u}^{\star}_{s,ij}(\omega)$ is the complex conjugate of $\dot{u}_{s,ij}(\omega)$.

Finally, the structural velocity variance may be obtained from

$$\sigma^2_{\dot{u}_{s,ij}} = \int_0^{\infty} S_{ss,ij}(\omega)d\omega \tag{B.30}$$

Structural velocity - particle velocity covariance: The structural velocity-particle velocity covariance $\sigma^2_{\dot{u}_{s,ij}\dot{u}_{p,i}}$ (= Cov[$\dot{u}_{s,ij}$, $\dot{u}_{p,i}$]) is the area under the structural velocity-particle cross-spectrum and is given by

$$\sigma^2_{\dot{u}_{s,ij}\dot{u}_{p,i}} = \int_0^{\infty} \dot{u}_{s,ij}(\omega)\,\dot{u}^{\star}_{p,i}(\omega)\,d\omega \tag{B.31}$$

Structure-particle relative velocity variance: The relative velocity between the structure and the water particles at the level of mass i and for vibration mode j is

$$\dot{r}_{ij} = \dot{u}_{p,i} - \dot{u}_{s,ij} \tag{B.32}$$

Assuming that $E[\dot{u}_{p,i}] = E[\dot{u}_{s,ij}] = 0$,

$$Var[\dot{r}_{ij}] = E[(\dot{u}_{p,i} - \dot{u}_{s,ij})^2] \tag{B.33}$$

giving

$$\sigma^2_{\dot{r},i} = \sigma^2_{\dot{u}_{p,i}} + \sigma^2_{\dot{u}_{s,ij}} - 2\sigma^2_{\dot{u}_{s,ij}\dot{u}_{p,i}} \tag{B.34}$$

$\sigma_{\dot{r}(i)}$ obtained from the above may then be re-substituted in equation (B.6) to obtain a better eatimate of the linearised fluid damping terms.

5. MULTIPLE PILES

The preceding theory applies to a wave force acting on a single vertical member or pile, or to a braced frame lying parallel to the wave front. However, if the structure consists of two or more members or frames, the wave forces acting on the separate parts of the structure are different, and depend in part on the ratio of wave length to the distance between the frames. The concept of a multiple pile transfer function [B.1] can be used to modify the drag and inertia coefficients to allow for this effect. This approach takes account of the relative positions of the members with respect to wave length at each component of frequency. The modified coefficients c_a' and c_b' are given by

$$c_a' = c_a\left(\sum_{m=1}^{n'} \sum_{n=1}^{n'} \cos k(x_n - x_m)\right)^{\frac{1}{2}} \tag{B.35}$$

$$c_b' = c_b\left(\sum_{m=1}^{n'} \sum_{n=1}^{n'} \cos k(x_n - x_m)\right)^{\frac{1}{2}} \tag{B.36}$$

where $(x_n - x_m)$ is the horizontal distance between the nth and m^{th} pile, k is the wave number (see page 216), n′ is the number of piles or frames in the direction normal to the wave front, and c_a and c_b can be obtained from the equations on page 256.

In practical applications, it is clear that there may be considerable uncertainty in the total force acting on the structure due to shielding effects, wave slam, the effects of members inclined to the vertical, but at present these influences cannot be allowed for in a rigorous way. It is suggested, however, that some account of this is taken when modelling Morison's coefficients C_d and C_m.

6. COMPUTATIONAL PROCEDURE

The calculations described above require considerable computational effort - particularly when they are used within an iterative level 2 reliability analysis. In particular, the areas under the various spectra need to be evaluated by numerical integration, and careful selection of frequency points is required to obtain sufficient accuracy without undue loss of computational efficiency.

BIBLIOGRAPHY

[B.1] Borgman, L.E.: *The Spectral Density for Ocean Wave Forces.* Coastal Eng. Conf., ASCE, 1965.

[B.2] Newland, D. E.: *An Introduction to Random Vibrations and Spectral Analysis.* Longman, London, 1975.

[B.3] Malhotra, A. K. and Penzien, J.: *Non-Deterministic Analysis of Offshore Structures.* Journal Eng. Mech. Div., ASCE, EM6, Dec. 1970.

[B.4] Lin, Y. K.: *Probabilistic Theory of Structural Dynamics.* McGraw-Hill, N. Y., 1967.

INDEX

GPSR Compliance

The European Union's (EU) General Product Safety Regulation (GPSR) is a set of rules that requires consumer products to be safe and our obligations to ensure this.

If you have any concerns about our products, you can contact us on ProductSafety@springernature.com

In case Publisher is established outside the EU, the EU authorized representative is:

Springer Nature Customer Service Center GmbH
Europaplatz 3
69115 Heidelberg, Germany

Batch number: 10683525

Printed by Printforce, the Netherlands